Radioguided Surgery

VADEMECUM

Radioguided Surgery

Eric D. Whitman, M.D.
Director, The Melanoma Center
Saint Louis, Missouri, U.S.A.

Douglas Reintgen, M.D.
Program Leader, Cutaneous Oncology
H. Lee Moffitt Cancer Center
and
Professor of Surgery
University of South Florida
Tampa, Florida, U.S.A.

CRC Press
Taylor & Francis Group
Boca Raton London New York

CRC Press is an imprint of the
Taylor & Francis Group, an **informa** business

VADEMECUM
Radioguided Surgery

CRC Press
Taylor & Francis Group
6000 Broken Sound Parkway NW, Suite 300
Boca Raton, FL 33487-2742
First issued in paperback 2019

© 1999 by Taylor & Francis Group, LLC
CRC Press is an imprint of Taylor & Francis Group, an Informa business

No claim to original U.S. Government works

ISBN 13: 978-1-57059-569-1 (pbk)

Visit the Taylor & Francis Web site at
http://www.taylorandfrancis.com

and the CRC Press Web site at
http://www.crcpress.com

Development of this book was supported in part by an unrestricted educational grant from U.S. Surgical Corporation.

Library of Congress Cataloging-in-Publication Data

Radioguided surgery / [edited by] Eric D. Whitman, Douglas Reintgen.
    p.   cm. -- (Vademecum)
Includes bibliographical references and index.
ISBN 1-57059-569-0
1. Breast--Cancer--Radiotherapy. 2. Breast--Cancer--Surgery. 3. Lymph nodes--Diseases--Diagnosis. I. Whitman, Eric D. II. Reintgen, Douglas Scott. III. Series.
[DNLM: 1. Surgical Procedures, Operative. 2. Radionuclide Imaging. WO 500 R129 1999]
RD667.5.R24 1999
617'.05--dc21
DNLM/DLC   98-53305
for Library of Congress   CIP

Contents

Editors

Eric D. Whitman, M.D.
Director, The Melanoma Center
St. Louis, Missouri, U.S.A.
Chapters 1, 3

Douglas Reintgen, M.D.
Program Leader, Cutaneous Oncology
Moffitt Cancer Center
Professor of Surgery
University of South Florida
Tampa, Florida, U.S.A.
Chapter 5

Contributors

Claudia G. Berman, M.D.
Associate Professor of Radiology
University of South Florida
Tampa, Florida, U.S.A.
Chapter 8

Charles E. Cox, M.D.
Professor of Surgery
University of South Florida
Program Leader
Comprehensive Breast Cancer Center
Moffitt Cancer Center
Tampa, Florida, U.S.A.
Chapter 6

C. Wayne Cruse, M.D.
Professor of Surgery
Division of Plastic Surgery
University of South Forida
Tampa, Florida, U.S.A.
Chapter 12

James V. Fiorica, M.D.
Professor, Gynecologic Oncology
H. Lee Moffitt Cancer Center
University of South Florida
Tampa, Florida, U.S.A.
Chapter 13

Jeffrey E. Gershenwald, M.D.
Assistant Professor of Surgery
Department of Surgical Oncology
The University of Texas M. D. Anderson
   Cancer Center
Houston, Texas
Chapter 4

Armando E. Giuliano, M.D.
Chief of Surgical Oncology
John Wayne Cancer Institute
Santa Monica, California, U.S.A.
Chapter 7

Edward C. Grendys, Jr., M.D.
Assistant Professor
Gynecologic Oncology
H. Lee Moffitt Cancer Center
University of South Florida
Tampa, Florida, U.S.A.
Chapter 13

Fadi F. Haddad, M.D.
Clinical Research Fellow
Cutaneous Oncology Program
Moffitt Cancer Center
University of South Florida
Tampa, Florida, U.S.A.
Chapter 5

Philip I. Haigh, M.D., FRCS(C)
Surgical Oncology Fellow
John Wayne Cancer Institute
Santa Monica, California, U.S.A.
Chapter 7

David A. Hillier, M.D.- Ph.D.
Instructor of Radiology, Washington
  University School of Medicine
Mallinckrodt Institute of Radiology
St. Louis, Missouri, U.S.A.
Chapter 2

Ni Ni K. Ku, M.D.
Associate Professor of Pathology
H. Lee Moffitt Cancer Center
University of South Florida
Tampa, Florida, U.S.A.
Chapter 9

Jane L. Messina, M.D.
Associate Professor of Pathology
University of South Florida
Tampa, Florida, U.S.A.
Chapter 10

James Norman, M.D.
Associate Professor of Surgery
  and Internal Medicine
Director of Endocrine Surgery
University of South Florida
Tampa, Florida, U.S.A.
Chapter 11

Lary A. Robinson, M.D.
Director, Division of Cardiovascular
  and Thoracic Surgery
Thoracic Oncology Program
H. Lee Moffitt Cancer Center
University of South Florida
Tampa, Florida, U.S.A.
Chapter 14

Merrick I. Ross, M.D., F.A.C.S.
Associate Professor of Surgery
Chief, Melanoma/Sarcoma Section
Department of Surgical Oncology
The University of Texas M. D. Anderson
  Cancer Center
Houston, Texas
Chapter 4

Henry D. Royal, M.D.
Professor of Radiology, Washington
  University School of Medicine
Associate Director,
  Division of Nuclear Medicine,
  Mallinckrodt Institute of Radiology
St. Louis, Missouri, U.S.A.
Chapter 2

Preface

Radioguided surgical procedures, including sentinel node mapping and biopsy, are currently taking the American surgical community by storm. The explosive yet poorly measured growth in these techniques is perhaps best indicated by the proliferation of clinical and basic science research papers presented on the topic at meetings such as the American College of Surgeons and the Society of Surgical Oncology.

In preparing this handbook, we have endeavored to have experts in their fields describe exactly what they do, how they do it, and why. Our goal is to create an indispensable resource for the practicing clinician: the ultimate arbiter of technical and logistical questions about radioguided surgical procedures. We have tried to avoid overly academic or review-oriented descriptions of the results from clinical centers of innovation. We prefer instead to focus on the future practitioners of sentinel node biopsy and other techniques: the practicing surgeon and other clinicians who will continue to provide the majority of clinical care for breast cancer and other diseases and who therefore must be able to safely and effectively perform these procedures for their patients.

We hope that you find this handbook essential to your clinical practice, and we invite your comments, suggestions, or questions.

Eric D. Whitman
Douglas Reintgen

Logistics and Organizational Aspects of Radioguided Surgery Programs

Eric D. Whitman

INTRODUCTION

Initiating a radioguided surgical program involves much more than just learning the technical aspects of the procedure(s) and purchasing an intraoperative gamma probe. Upon review of some of the major publications in the field,[1-5] I believe there are certain nonscientific issues that stand out as essential, yet behind-the-scenes; the *infrastructure* of successful radioguided surgical (RGS) programs. For example, patient scheduling for procedures is immensely more complicated, because it will routinely involve the coordination of multiple departments within the hospital, including the operating room, nuclear medicine, and pathology. This coordination must function smoothly and precisely to enable the procedure to be performed within the ideal window after radionuclide injection, with timely pathologic evaluation (which likely involves specialized personnel). The pathologic evaluation itself is critical; the data for sentinel node biopsy is predicated upon the *lowest possible false negative rate* when compared to the pathology of the remainder of the nodal chain. If you cannot ensure that you are taking all possible steps to minimize the false negative rate, your institution should not be performing sentinel node biopsy, since you are ultimately doing a disservice to the patient because you are not providing the same quality information as that published in the medical literature. In order to achieve this, a comprehensive

Radioguided Surgery, edited by Eric D. Whitman and Douglas Reintgen.
© 1999 Landes Bioscience

and consistent pathology evaluation algorithm employing the techniques described elsewhere in this handbook must be implemented.

Many of the recommendations in this chapter cannot unfortunately be based on published guidelines; rather they have been synthesized by me during my experience as the organizer of radioguided surgical programs at two philosophically and geographically distinct hospitals, the first a large urban university medical center and the second a medium sized suburban community hospital. This chapter, therefore, focuses on the logistics behind a successfully implemented radioguided surgical program.

ORGANIZATIONAL ISSUES

The most important organizational issues that I have encountered in my involvement with RGS programs at two very different institutions are the need to develop a consistent set of standardized procedures for all components of the program, and to have agreement between all involved parties (i.e., pathology, nuclear medicine, surgery, administration, etc.) on how these policies and procedures will be implemented prior to scheduling the first sentinel node case. These organizational issues can be divided into Program Initiation, Nuclear Medicine, Pathology, and Scheduling components.

INITIATING A PROGRAM

In most institutions, surgeons have been the driving force behind initiating these procedures, usually after being trained at one of the specialized courses. As discussed in another chapter in this handbook, many courses do not appear to adequately prepare attendees for the logistical hurdles they must negotiate upon their return to their home institutions.

This begins with surgeon training and credentialing. The hospital, with input from interested physicians, should decide *before* any RGS is performed who would be allowed to perform what procedures, and in which situations. This ideal situation is often not possible for a variety of reasons. Alternative solutions are discussed in chapter 3 ("Training and Credentialing"). The important point is that some regulation of RGS should initially occur to prevent inappropriate usage of the advanced technology in ways that might ultimately harm patients. These regulations should include training requirements, credentialing, and outcome measurements and analysis, to ensure that the patients are receiving a level of care consistent with published results from more experienced referral centers.

The involved physicians must also decide which RGS procedures will be performed and whether these procedures are best performed under a research protocol with Institutional Review Board (IRB) approval. The decision to seek IRB approval for some or all of the RGS operations is to an extent dependent upon the comfort and training level of the individual surgeon, the clinical research interests of the involved physicians, and the specific requirements of the institution. Most institutions today are *not* obtaining IRB approval for sentinel node biopsy for melanoma, unless there are specific research interests being pursued. Although there is an ongoing clinical study evaluating SLN biopsy for melanoma coordinated through the John Wayne Cancer Institute,[6] there have been other publications and programs that strongly state that SLN biopsy for melanoma is the standard of care.[7,8] Unlike breast cancer, standard surgical treatment of melanoma does not include lymph node dissection, despite multiple attempts to justify "prophylactic" lymph node dissection in recent years.[9] However, the recent approval of interferon alfa-2b (Intron A®, Schering Plough Inc., Madison, New Jersey) to treat patients at high risk for recurrent melanoma, especially those with lymph node positive disease, has put a premium on the early detection of nodal metastases in patients with at least intermediate thickness melanoma. The highest level of sensitivity to detect nodal disease is currently available only through SLN sampling, provided the pathologic examination meets the criteria discussed elsewhere in this chapter and others. Previously, there was no treatment that was proven effective for patients at high risk of melanoma recurrence, but a long-term multicenter randomized prospective trial (ECOG 1684) established Intron A as the only approved adjuvant therapy for patients at risk.[10] Thus, SLN biopsy for melanoma detects nodal dissemination at an early (clinically occult) time point, while enabling the separation of those patients whose high-risk primary lesions have apparently not yet metastasized. Further, SLN biopsy eliminates the logistical need for prophylactic lymph node dissections, converting all lymph node dissections in melanoma patients to "therapeutic." For these reasons, and the relative ease with which most of the melanoma SLN are located, IRB approval is not generally pursued by most physicians performing SLN biopsy for patients with this cancer.

The situation is much different for SLN biopsy for breast cancer, a disease far more prevalent in the United States than melanoma. Due to its more common occurrence, it is more likely to be treated by nonspecialist physicians in a community setting than melanoma. Conversely, many melanoma patients in the United States and elsewhere are referred relatively early in their treatment to specialists in the field, who will also be more experienced with SLN biopsy techniques. Considering the differences in disease prevalence and typical referral patterns in this country, most potential RGS procedures for general surgeons are likely to be for patients with newly diagnosed breast cancer. This scenario presents several problems. First, the technical aspects of SLN biopsy for breast cancer are generally considered to be more difficult than the same procedure for melanoma. The reasons for this difference are covered elsewhere in the handbook. Second, there are fewer published data regarding SLN biopsy for breast cancer than for melanoma, and the data published reflect the extended learning curve and greater potential for failure associated with SLN biopsy for breast cancer. Third, breast cancer care in the United States occurs within a politically charged environment, amplifying any mistakes while simultaneously encouraging rapid adoption of any technique offering potential advantages.

I strongly believe that all breast cancer SLN procedures should at least initially be done under an IRB approved protocol at this time (early 1999). An ideal single-armed protocol should specify that all patients would undergo SLN biopsy followed by standard axillary node dissection, regardless of the pathology of the sentinel node. The purposes, in my mind, of performing all SLN biopsies for breast cancer under IRB approval are:

1) to ensure that the sentinel node biopsies are performed with consistent technique by surgeons trained in these procedures,

2) to provide consistent nuclear medicine and pathologic examination protocols and algorithms for all breast cancer SLN biopsies,

3) to comprehensively monitor clinical outcomes from the procedures, *and analyze these outcomes to **prove statistically significant equivalence** between your institution's results and those published in the literature from major referral centers.*

Thus, the ideal IRB approved protocol will enable the investigator to state at its predetermined accrual endpoint that SLN biopsy for breast cancer at that institution, performed by the surgical co-investigators and processed by the nuclear medicine physicians and pathologists according to the criteria and algorithms specified to the IRB, is statistically equivalent to SLN biopsy techniques and results published in the literature. Further, based on the high quality outcomes obtained, that particular institution and clinicians are now able to offer SLN biopsy without routine axillary node dissection, as the procedure has evolved at more experienced centers.

The American College of Surgeons has obtained National Cancer Institute funding for clinical trials in surgical oncology (ACSCOG program).[6] One of the first trials will be a multicenter examination of sentinel node biopsy for breast cancer. As with any large trial, there will be pros and cons to participating, and most hospitals/physicians will not. This is not meant to make any statement regarding that trial; that is simply how these situations typically evolve. Nonetheless, the positive media reports of sentinel node biopsy for breast cancer will only accelerate, placing significant pressure on all physicians treating this disease to offer SLN biopsy sooner rather than later. In this setting, I have written a complete, generic, non-institution specific, protocol for breast cancer SLN biopsy, meeting the criteria and schematic examples set forth above. This protocol and consent form can be obtained free of charge by e-mail from me at e_whitman@hotmail.com. This protocol is designed so that it may be submitted *as is* to any IRB after the name of the principal and any associate investigators have been typed onto the front page. This protocol is not designed to be a replacement for the ACSCOG trial; it is merely a way for all institutions and physicians to safely and consistently begin to perform SLN biopsy for breast cancer.

Performance of other RGS procedures, particularly minimally invasive radioguided parathyroidectomy, without IRB approved protocol monitoring is less well defined, as these procedures to date have been less commonly utilized. However, the decision to perform any RGS procedure with IRB approval must only be made with the recognition that part of what IRB approval provides is medicolegal protection, by ensuring that all participants sign a detailed informed consent and understand the investigational nature of their therapy. Performing any RGS procedure without an investigational protocol exposes the clinician to this medicolegal risk, and demands full disclosure to the patient of the procedural components, published literature, personal clinician experience, and risk-benefit comparison.

NUCLEAR MEDICINE

Lymphoscintigraphy for SLN procedures is performed using standard techniques and protocols, well described in chapter 8. Most radiologists are at minimum familiar with, if not experienced in, this procedure and the technical aspects of lymphoscintigraphy itself should be expected to go well. There are, however, several logistical issues to be considered from the perspective of coordinating nuclear medicine activities with other clinical specialties within a RGS program.

First, it is important to discuss the purposes of lymphoscintigraphy with the nuclear medicine physicians before initiating the RGS program. Unlike a standard lymphoscintigram, a study performed on the day of SLN surgery must meet several constraints, mostly to facilitate scheduling by the surgeon and operating room. As discussed in a subsequent section of this chapter, the schedule a patient follows on the day of SLN surgery requires multiple departments within the hospital to function within fixed time constraints. This type of scheduling is atypical for most nuclear medicine studies. We have had to reinforce to the nuclear medicine physicians, technologists, and administrative staff our scheduling priorities.

From the surgical standpoint, the essential elements of the sentinel lymph node lymphoscintigraphy are:

1) identify where the sentinel node(s) is/are (*basin only*, see below);
2) scan all basins at risk to confirm which do *not* have SLN within them; and,
3) deliver the patient to the operating room holding area with copies of the relevant scans so as to not delay the start of the scheduled operation.

We do not ask that the radiologists mark or tattoo the skin over the SLN, unlike many other programs. We have found that this requires extra time in the nuclear medicine suite and also unfortunately may be misleading to the surgeon. The position of the patient during lymphoscintigraphy is likely to be different than the position during the operation, causing the skin marking to move relative to the underlying SLN once the patient is fully positioned, prepped and draped. I have personally made (and have seen others make) misplaced incisions because of well intentioned but ultimately inaccurate skin markings. Also, marking the skin location of the SLN within a standard, routinely dissected lymphatic basin such as the axillary or inguinal locations is superfluous; all surgeons should be experienced enough to locate an axillary sentinel node without the assistance of a skin mark. With the patient completely positioned for the procedure on the operating room table, I make my own skin marks, using the intraoperative gamma probe to interrogate the skin overlying the nodal basin identified by the lymphoscintigraphy, before prepping and draping the area. This is particularly important in areas other than the axilla or groin, such as the head and neck, flank, and scapula, where both lymph node location and patient positioning are more variable.

Breast SLN lymphoscintigraphy is slightly different, as the melanoma SLN tend to "light up" more quickly and with more intensity, enabling melanoma patients to complete their scans sooner. Breast cancers have a more predictable drainage pattern, so that typically only one view is necessary, of the ipsilateral chest and axilla. The breast SLN tend not to concentrate as much radioactivity, which may make them more difficult to locate on lymphoscintigraphy, particularly in patients with upper outer quadrant lesions, where proximity to the high concentration of radionuclide at the primary site may obscure the SLN on the scan.

Early in our experience, we had significant problems scheduling breast SLN lymphoscintigraphy and surgery, due to difficulties obtaining satisfactory (using standard, non-SLN criteria) images which led to large delays in transporting the patient to the operating room. We have discussed these issues at length with our nuclear medicine physician colleagues and have arrived at compromises designed to enable high quality lymphoscintigraphy to coexist with the scheduling constraints of the surgeon and the operating room. Following injection of the radionuclide at the primary site in the breast, the patient is scanned once, at about 75 minutes. The scanned area encompasses the ipsilateral chest, showing the axilla, supraclavicular, and internal mammary nodal chains. The patient and a copy of the scan, regardless of the findings, are then sent to the operating room. We have found (unpublished data) that nodes are unlikely to show up on scans if they do not appear by this time point. Further, we utilize both the blue dye and the radionuclide to localize the breast SLN and have found these techniques complementary, *particularly* in cases where no lymph nodes are visualized by lymphoscintigraphy. In our experience (unpublished data), SLN can still be localized following a "negative" lymphoscintigram, either by the blue dye alone and/or with the gamma probe in cases where the primary injection site obscures the view of the axilla by the scanner. During the time between radionuclide injection and scans, our breast SLN patients are transported to admitting or preoperative holding areas, as necessary, to complete their pre-surgical evaluation and registration, more effectively utilizing their time in the hospital preoperatively.

For minimally invasive parathyroidectomy, we have asked our radiologists to follow the protocol described by the group at the University of South Florida, available on the internet at http://endocrineweb.com. We have emphasized the timing issues, as described by Dr. Norman,[11] where the maximal separation of counts in the parathyroid and thyroid glands should be between 2 and 4 hours. It is, therefore, imperative that the patient's procedure begin somewhere in this time frame.

Overall, we have found the nuclear medicine physicians very responsive to our needs, regardless of the institution involved. However, it has been very rewarding for us and our patients to establish the operative and scheduling constraints of lymphoscintigraphy and RGS with them, so that the scans are performed in a consistent manner that maximizes the information provided to the surgeon while minimizing potential disruption of surgical schedule and acquisition of unnecessary data.

PATHOLOGY

The use of sentinel lymph node biopsy for cancer staging is based on the hypothesis, subsequently supported by data from multiple centers in several different cancers, that the SLN is **the most informative node** for that primary cancer; **that if the cancer has regionally metastasized, it will be to that node first,** and finally, **if the SLN is not pathologically positive, no other lymph node in the patient can have any evidence of metastasis.** The pathologic examination of the SLN may be the key component testing this hypothesis, to ensure that adequate evaluation of the sentinel nodes are performed, while proceeding appropriately and cost effectively in this era of cost containment in medical care. This area has proved to be a struggle in many centers, as there is no absolute proof that earlier detection of cancer metastases, particularly at the microscopic level, contributes to patient prognosis in melanoma or breast cancer. Concerns have also been raised as to whether a pathologist/institution can be reimbursed for the extra time, reagents, and testing necessary to perform serial sections and immunohistochemical stains on some or all of the SLN harvested for each patient. Our SLN program was initially encumbered by these same concerns, but upon review of the literature it was clear that all of the leading centers of innovation utilized an aggressive pathologic evaluation scheme, involving both serial section analysis and immunohistochemical staining. If we did not implement a similar pathologic algorithm, would we be providing the same level of prognostic information to the patient or referring physician?

I concluded, based on discussions with physicians at the pioneering centers in RGS, that it would be impossible for us to have confidence in our SLN pathology results without performing serial section and immunohistochemical analysis, since we would likely miss micrometastatic disease, increasing our regional recurrence risk. This conclusion has been reinforced by several recent publications. In the first instance, all melanoma patients with recurrence in the nodal basin after negative SLN biopsy had the tissue blocks reexamined with serial section and immunohistochemical analysis. An overwhelming percentage of the cases would have been pathologically positive if the more comprehensive/aggressive approach had been taken initially.[12] Another report confirmed the utility of a more comprehensive pathologic examination, concluding that about 40% of patients with micrometastatic disease would be "missed" without serial sections and immunohistochemical staining (see chapter 10). Finally, researchers at John Wayne Cancer Institute examined the validity of the more comprehensive pathologic algorithm to examine SLN in breast cancer by also subjecting the non-SLN to this aggressive pathologic examination protocol. In 1087 non-SLN, only 1 node had evidence of micrometastatic disease not found on routine bivalved H&E examination.[13] This underscores both the need for serial sections and immunohistochemical staining

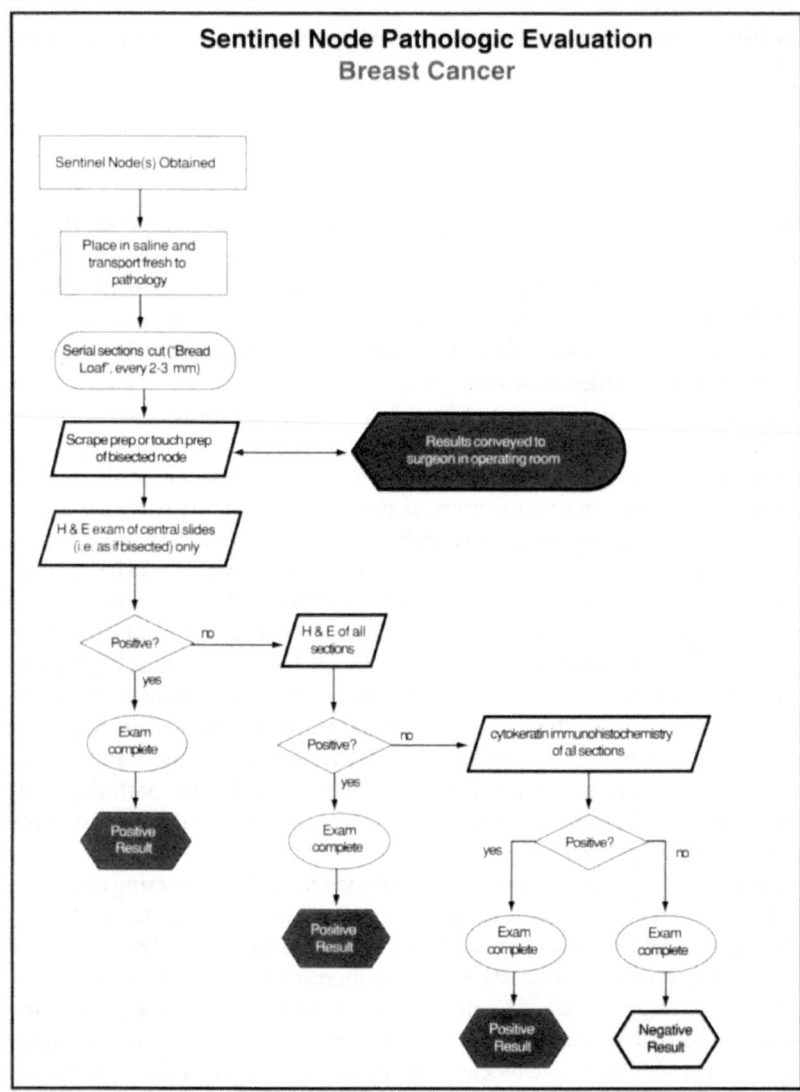

Fig. 1.1. Sentinel node pathologic evaluation.

of the SLN and the apparent safety of *not* performing as comprehensive an examination of the non-SLN.

During implementation of a single-armed IRB approved study of SLN biopsy for breast cancer at our institution, we included an algorithm for comprehensive pathologic examination of the SLN, shown in Figure 1.1. This algorithm was developed by myself and Dr. Charles Short, director of pathology at Missouri Baptist Medical Center in St. Louis, Missouri. The algorithm is designed to **provide**

up to four separate data points for each SLN. Each SLN is separately examined and reported upon at each level, but the rules for "stopping" further examination apply if any SLN is positive, to eliminate unnecessary examinations and expense. The first data point is an intraoperative examination, using either touch prep or scrape prep techniques of the bivalved central section of each SLN, as described in chapter 9. The second data point is H&E examination of the bivalved or central sections of the SLNs, as would otherwise be performed for all non-SLN biopsies. All SLN will undergo both of these examinations, regardless of the results of the intraoperative (immediate) testing. However, should any node be positive by the second (permanent H&E slide) test, no further pathologic examinations are warranted. This avoids the hours and expense of performing serial section analysis and immunohistochemical staining on all SLN in every case. We also do not rely solely on the immediate, intraoperative evaluation, since we are concerned about false positive results by this method, particularly since not all pathologists will at least initially be as experienced as Drs. Ku (chapter 9) or Messina (chapter 10) in these techniques. Should the immediate examination be positive for metastatic disease, but the central H&E section is negative, we will proceed on to serial section analysis, and if necessary, immunohistochemical staining to either confirm or refute the immediate findings from the touch or scrape prep. For these reasons also we are not yet using the immediate intraoperative evaluation to change our surgical plan until we have confirmed the validity of these techniques in our pathologist's hands. Therefore, all melanoma patients will have SLN biopsy performed initially without intraoperatively changing to a complete node dissection unless the node is grossly involved with malignancy. For breast patients, we perform the SLN biopsy as part of an IRB approved protocol as above, so that all patients currently receive complete axillary node dissection immediately after the SLN biopsy. We anticipate acquiring enough data during this IRB approved breast SLN protocol to be able to fully implement immediate or intraoperative examination using the techniques described elsewhere when we begin breast SLN biopsies off-protocol.

The third data point is the result of H&E examination of serial sections, prepared by slicing the SLN at 2-3 mm intervals and preparing slides for examination at each level. It is anticipated that 20-30% of negative SLN by central section examination will be positive by this level of testing. Finally, if all previous tests have proved negative (or if the only positive test result was the immediate evaluation), the serial sections are submitted for immunohistochemical analysis, using the antibodies described elsewhere in this handbook by Drs. Messina and Ku, respectively. Each data point of the pathologic examination process is recorded separately, as shown on our breast SLN data sheet (Fig. 1.2), which is also included in the document set of our IRB protocol.

Implementation of this standardized pathologic algorithm, developed jointly by the surgical and pathologic teams at our institution, has been very successful clinically and administratively, involving both departments in the development of a new way of approaching lymph node biopsy and examination. I strongly recommend a similar approach at institutions new to SLN biopsy, to enable from

**Breast Sentinel Lymph Node Mapping Study
Data Entry Sheet**

v2.2   edw, 1998

Patient Code Number

Information about Primary

| Pathology | ☐ Ductal   ☐ DCIS   ☐ Other...   ☐ Lobular   ☐ Tubular | Check all that apply |

Side   ○ Right   ○ Left

Location   ☐ Upper   ☐ Outer   ☐ Central   ☐ Lower   ☐ Inner   Check all that apply

Size (mm)   by mammography   by clinical exam (if applicable)   by pathology

Mode of diagnosis   ☐ FNA   ☐ Surgical Excision   ☐ Core bx   ☐ Surgical Incision   Check all that apply

Lymphoscintigraphy

format hh:mm (24 hr)
Injection Time
Incision Time

SLN site (on scans)   ☐ Axilla   ☐ Supraclavicular   ☐ Internal mammary   ☐   None   Check all that apply

Time since Tc99 injection

Surgery

Op Date   **Age (at surgery)**

Surgeon   **Hospital**

Localization Technique (for Tc99m injection)   ☐ Palpable   ☐ Mammogram   ☐ Other...   ☐ Ultrasound   ☐ Skin injected   Check all that apply

Surgical Procedure   ☐ Lumpectomy   ☐ Quadrantectomy   ☐ Mastectomy   ☐ Needle localized   Check all that apply

Gamma Counter   ○ Navigator   ○ Neoprobe   ○ C-Trak

Blue Dye Used?   ○ Yes   ○ No

CPS
(optional values)   **Background**   **non-SLN**   **Wound after Excision**

SLN Pathology

	Nodes	cps	SLN blue?	Ax Node Level	Immed	Bisect	Serial	IHC	
SLN			○ Yes ○ No	○ I ○ I ○ II					Pathology:
Number harvested			○ Yes ○ No	○ I ○ I ○ II					Pos (positive)
			○ Yes ○ No	○ I ○ I ○ II					Neg (negative)
			○ Yes ○ No	○ I ○ I ○ II					ND (not done)
			○ Yes ○ No	○ I ○ I ○ II					
			○ Yes ○ No	○ I ○ I ○ II					

non-SLN   **Number in specimen**   **Number positive**

Signature of person completing form   Date

Fig. 1.2. Breast sentinel lymph node mapping study data entry sheet.

the beginning a pathologic protocol that provides results consistent with those published by the leading centers of SLN innovation.

SCHEDULING A PROCEDURE

To schedule a RGS procedure, it is necessary to coordinate the availability and timing of several different procedures and departments within the hospital. We have found the following guidelines to work best:

1) **We never obtain a radionuclide localization study before the day of surgery,** except initially for patients with primary hyperparathyroidism as we and our referring physicians became more comfortable with that procedure. We found that physicians preferred to obtain a Sestamibi scan before referring the patient for surgery, as this is a new procedure and they liked to know before the actual referral whether a MIRP was possible. Following the initial cases, preoperative Sestamibi scans are more commonly done only on the day of surgery.

For melanoma and breast cancer cases, there is no reason to subject the patient to the added discomfort, time, risk, and cost of lymphoscintigraphy at any time except immediately before surgery. For these two diseases, the decision to perform SLN mapping is not based on the results of the lymphoscintigram, but on the patient's diagnosis, clinical stage, and risk of nodal metastases. Regardless of the results of a lymphoscintigram performed at a time remote from the RGS procedure, another radionuclide injection will need to be accompanied by at least one set of scans, so that the radiologist and hospital can be reimbursed for the injection. Therefore, even if the nodal chain has been "localized" by a lymphoscintigram performed days or weeks in advance, a second scan will still be necessary on the day of surgery. For melanoma cases, the lymphoscintigraphy will without fail show a sentinel node or nodes somewhere; the decision on patient positioning or sometimes even level of anesthesia will depend upon the findings of the scan and for this reason we ask our anesthesiologists and operating room staff to be flexible in their preparation for melanoma SLN cases, as we never know exactly how we will approach the case until after the scans are complete. In our experience, breast cancers will occasionally not show a definite SLN on preoperative lymphoscintigraphy, even though one can often be found at the time of surgery, even in the absence of a "positive" lymphoscintigram. This points out the complementary utility of the lymphazurin blue dye and the radionuclide, as discussed elsewhere.

2) **Schedule all phases of the RGS procedure through one knowledgeable person** to try to ensure maximal coordination of the various departments. Our RGS scheduling is completely coordinated by a single RN experienced in surgical oncology, who understands the issues and constraints of both the surgery and the surgeon. A typical day of surgery schedule for our RGS patients is as follows:

0700	Report to admitting, NPO after midnight. Registration paperwork completed; blood tests and admitting radiographs (if not already completed) are done.
0815	Patient transported to nuclear medicine
0830	Radionuclide injection
0845-1000	Scans performed
1015	Patient transported to operating room holding area with scans
1030	Patient evaluated by anesthesia, surgeon reviews scans to finalize choice of anesthetic requirements and patient positioning.
1100	Scheduled case start time

We provide the patients and their families with an outline of their day of surgery schedule, so they understand the logistics of the day. We also prepare them for what may be a long day by emphasizing in our communications with them the complexity and uncertainty in coordinating several different departments of the hospital, and that they should be prepared to spend most of the day at the hospital, even though the surgery itself is outpatient.

3) **Prepare mentally for the inevitable scheduling breakdowns.** In addition to the patients and their families, the operating surgeons must accept the difficulties in getting all of these procedures and tests to be performed in all patients without delays. Some of my colleagues have been frustrated by the hospital's seeming inability to get a lymphoscintigraphy done in time so that the SLN procedure can be successfully fit into an already tight operating schedule. The following "rules" are helpful:

> *-Do not attempt to perform any RGS procedures before 11 AM* (noon for breast cancers) until your institution has experienced 15-20 cases.
>
> *- Do not schedule outpatient office hours after your first few SLN procedures,* until you and your office have a better appreciation for how long the surgery and its preceding schedule will take.
>
> *- Minimize the number of non-SLN cases you schedule after a SLN procedure,* to avoid delays to other elective surgeries.

4) **Be creative and prepared to modify your scheduling system based on what works best at your institution.** We are constantly tinkering with the logistics of our SLN scheduling. For example, we have tried to actually schedule patient transport aides, to improve the efficiency of the transport process. Another modification of our schedule has been to send the patient down to nuclear medicine for radionuclide injection *before* completing the hospital registration paperwork, etc., then bringing the patient back to the admitting area during the gap between injection and scanning. The results of these modifications are mixed; cases still seem to start at 1100 at the earliest.

Finally, it is important to remember that for melanoma SLN cases, the surgical approach and position may change based on the location and number of sentinel node basins identified by the preoperative lymphoscintigraphy. This is also possible to a lesser extent with breast SLN cases (supraclavicular and internal mammary sentinel nodes are possible). This is most likely to occur with head and neck melanomas or torso melanomas. For torso melanomas, we always request a bean bag cushion on the operating table, as the axillary nodal basins are often involved, and it is possible to put the patient in the decubitus position and allow access to both the SLN basin and the melanoma primary. Operating room personnel and anesthesiologists at this institution now expect that the final choice of anesthetic technique and patient position will not occur until after the scans have been reviewed by the surgeon.

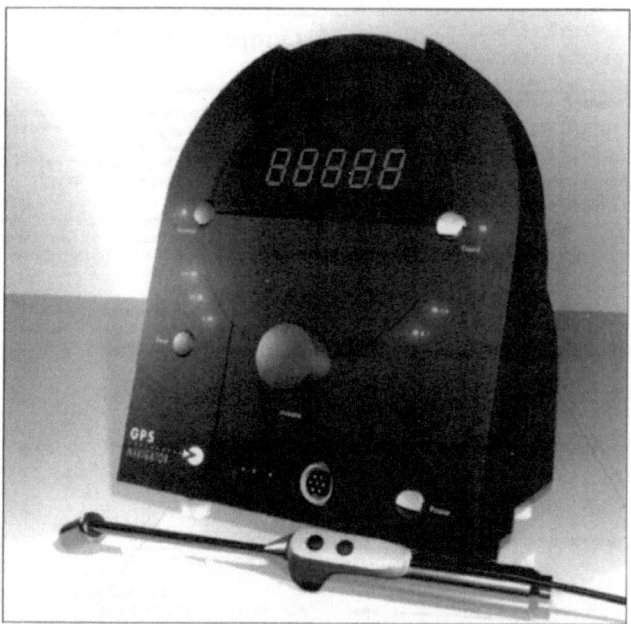

Fig. 1.3. (See Color Insert for color representation.) Example of an intraoperative gamma probe, specifically designed and optimally engineered for radioguided surgical procedures, the Navigator gamma guidance system (US Surgical Corporation, Norwalk, CT).

PURCHASING A PROBE

Before any RGS procedures can take place, it will obviously be necessary for your institution to purchase a gamma detector for intraoperative use. There are currently several different devices available on the marketplace, and it is not the intention of this author or textbook to recommend one over the other. Currently available gamma detectors are different in several ways, but market-driven changes and upgrades make it impossible to discuss the current models in a timely fashion within a handbook like this. I recommend evaluation of each probe by the involved surgeons, with attention to the three key components of any RGS device:

1) The ability to provide directional, or vectoring, information, to guide surgical dissection and removal of radiolabeled tissue, by use of collimation to narrow the "field of view" of the probe,

2) Elimination of extraneous or interfering radiation readings, from either the primary site or background radiation in the room or building, with adequate shielding of the radiation detector probe, and

3) Discrimination between radioactive and non-radioactive tissue, including filtering of radiation soft tissue scatter, through device incorporation of energy threshold adjustments.

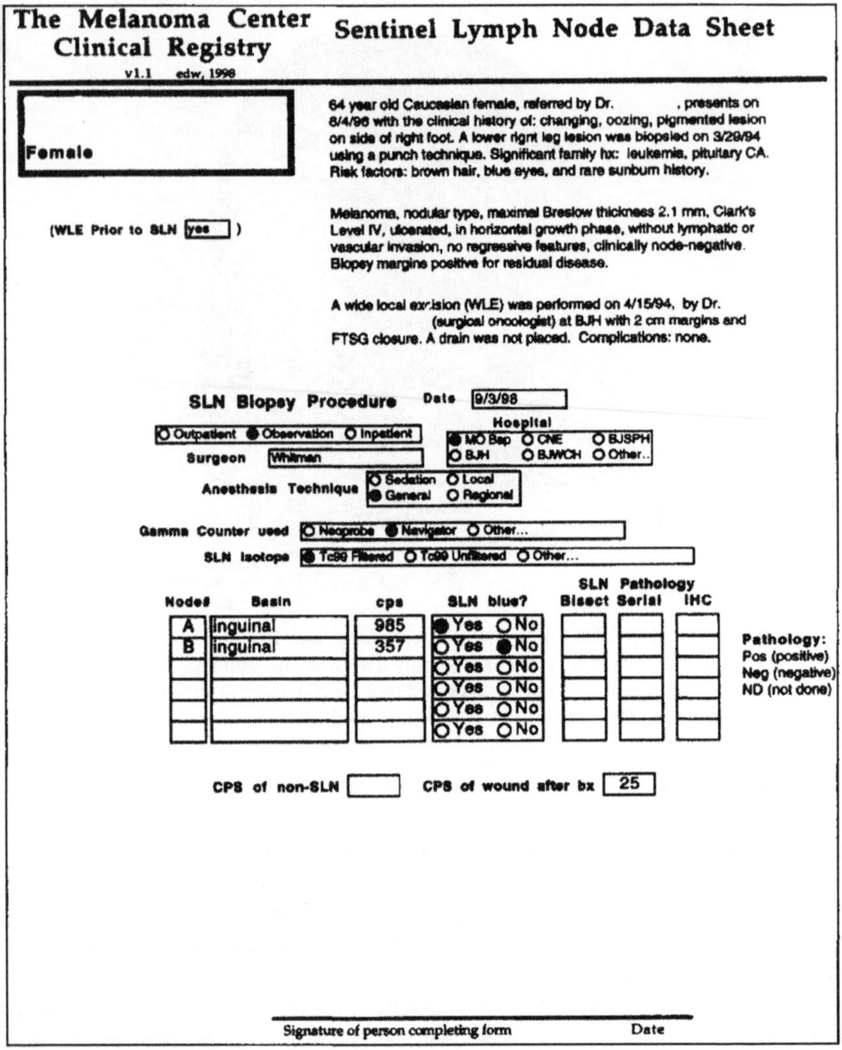

1

The Melanoma Center Clinical Registry

v1.1   edw, 1998

Sentinel Lymph Node Data Sheet

Female

64 year old Caucasian female, referred by Dr.   , presents on 8/4/96 with the clinical history of: changing, oozing, pigmented lesion on side of right foot. A lower right leg lesion was biopsied on 3/29/94 using a punch technique. Significant family hx: leukemia, pituitary CA. Risk factors: brown hair, blue eyes, and rare sunburn history.

(WLE Prior to SLN [yes])

Melanoma, nodular type, maximal Breslow thickness 2.1 mm, Clark's Level IV, ulcerated, in horizontal growth phase, without lymphatic or vascular invasion, no regressive features, clinically node-negative. Biopsy margins positive for residual disease.

A wide local excision (WLE) was performed on 4/15/94, by Dr.   (surgical oncologist) at BJH with 2 cm margins and FTSG closure. A drain was not placed. Complications: none.

SLN Biopsy Procedure   Date 9/3/98

Hospital

O Outpatient ● Observation O Inpatient   ● MO Bap   O CNE   O BJSPH
Surgeon Whitman   O BJH   O BJWCH   O Other..

Anesthesia Technique   O Sedation O Local
                           ● General   O Regional

Gamma Counter used   O Neoprobe ● Navigator O Other...

SLN Isotope   ● Tc99 Filtered O Tc99 Unfiltered O Other...

Nodes	Basin	cps	SLN blue?	SLN Pathology Bisect	Serial	IHC
A	inguinal	985	● Yes O No			
B	inguinal	357	O Yes ● No			
			O Yes O No			
			O Yes O No			
			O Yes O No			
			O Yes O No			

Pathology:
Pos (positive)
Neg (negative)
ND (not done)

CPS of non-SLN [   ]   **CPS of wound after bx** [25]

Signature of person completing form   Date

Fig. 1.4. Sentinel lymph node data sheet for melanoma procedures.

OUTCOMES MANAGEMENT

As your institution progresses in its development of a RGS program, it is imperative that appropriate outcomes be analyzed, to document the success rate of your efforts. In the near future, it is likely that both third party payers and patients will expect to see some objective evaluation of the volume and results of all surgical procedures, particularly newer ones or those perceived as requiring special skills and/or training. This evaluation is easiest to initiate at the beginning of a program, rather than in midstream.

Minimally Invasive Radioguided Parathyroidectomy (MIRP) Patient Information Registry

edw, 1998

Patient Code Number

3025

Demographics

DOB 8/22/1946

Preop labs   Calcium 16.9   Intact PTH 149

Sestamibi Scan

Patient Code 3025

Scan date 9/25/98

Head

RLL adenoma

Heart

Sestamibi Scan

format hh:mm (24 hr)   Preop Localization Results

Injection Time 13:00   ○ None

Incision Time 15:50   ☐ RUL   ☐ LUL

Time since Tc99 Injection 2:50   ☒ RLL   ☐ LLL
                                   ☐ Mediastinal

Check all
that apply

Surgery

Op Date	9/25/98	Age (at surgery)	52
Surgeon	Whitman	Hospital	MO Baptist

Gamma Counter ● Navigator ○ Neoprobe ○ C-Trak

Op Time (min) 43

Type of Surgery ☒ MIRP ☐ Std PTHectomy ☐ Thyroid lobectomy ☐ Other...

CPS

CPS / ratio to minimum

Timing	Probe	RUL		RLL		LUL		LLL	
Pre-inc	14 mm	392	1.03	389	1.02	392	1.03	382	1.00
Pre-inc	10 mm angled	56	1.08	53	1.02	52	1.00	52	1.00
Exposed	14 mm	378	1.00	705	1.87	395	1.04	379	1.00
Exposed	10 mm angled	48	1.04	112	2.43	46	1.00	55	1.20
Post-exc	10 mm angled	56	1.17	53	1.10	48	1.00	55	1.15

Adenoma cps 60   Central thyroid cps post exc 48   MIRP ratio 125

Final Pathology ● Adenoma ○ Hyperplasia ○ Not PTH ○ Other...

Comment Large adenoma removed.

_____   _____
Signature of person completing form   Date

Fig. 1.5. Minimally invasive radioguided parathyroidectomy (MIRP) patient information registry.

DATA ACQUISITION, STORAGE AND ANALYSIS

The following guidelines and suggestions are offered based only on our specific clinical experiences, and on my personal experience as a database programmer over the past 20 years. The data acquired by each institution is different and is governed by the needs and interests of the principal "investigators" or primary clinicians. Additionally, any national protocols such as the coming ACSCOG trial(s) tend to acquire more data to fit all the criteria and interests of the organizers. Based on my experience, I have found it best to attempt to streamline and reduce the amount of information gathered for data storage. Much of the data accumulated for routine studies ends up serving no useful purpose; it is never cited in publications, reports, or presentations, but was included initially because it was thought to be important and then never removed.[14] Data should be acquired and stored so that when we review our data registry we have a better understanding of how our RGS practice is evolving, the number and type of procedures we are performing, and what the results are.

Our primary data entry sheets are designed to fit on one page only and are disease specific. Our data sheets for breast SLN, melanoma SLN, and MIRP procedures are shown in Figures 1.2, 1.4, and 1.5, respectively. First, demographic information is stored to document what kind of patients had this procedure done. Second, diagnostic data is recorded, enough to classify patients by the extent of their disease and types of treatment given. Third, preoperative studies should be recorded, with their results. For RGS procedures, this is typically the nuclear medicine scan and possibly any preoperative staging studies.

Fourth, the operative data is recorded. The first three types of information are ideally recorded preoperatively, by either the physician or a specialized nurse clinician, but the operative data is most commonly recorded by the circulating nurse in the operating room. This process is facilitated by a simplified data recording sheet, with large and obvious data entry points, and minimal duties for the "recording nurse" other than actually writing down specific information. In that regard, I have eliminated much of the information that many other surgeons report recording. **Our only goals in recording information about sentinel nodes are to confirm the identify of the biopsied node as a sentinel node, using the criteria discussed elsewhere, and to subsequently prove that there are no further SLN in that basin.** Therefore, for each SLN, the only intraoperative data recorded are the identifier (which much be identical to the identifier sent with the specimen to pathology, usually either a number or a letter), the basin, the counts per second(cps) ex vivo, and whether or not it was blue-stained. After all SLN are removed, I record the cps of the basin after excision to document the absence of other SLN. Recording of non-SLN cps is usually superfluous, as the current devices available allow clear distinction of the radioactivity levels between sentinel and non-sentinel nodes. After excision, there are boxes for the pathologic findings for each level of pathologic examination of each SLN, as described earlier in this chapter. For breast SLN patients, the pathologic findings of the completion lymph node dissection are recorded on this same data sheet; node dissection results for melanoma patients are recorded elsewhere in our data registry (form not shown).

During MIRP procedures, our data is more focused on the localization information provided by the gamma detection device. We record the settings of the device, the type of probe used, and the cps in all four quadrants of the thyroid gland (right upper, right lower, left upper, and left lower) initially, after incision, and after thyroid gland dissection. Finally, the cps of the excised gland is recorded, and the ratio of the parathyroid tissue cps to the central thyroid gland (documented to be important by Dr. Norman's group).[11]

Our information is stored in computerized, password protected, data registries running on Filemaker Pro v4.0 for Windows/Macintosh (Claris Corp., Cupertino, California). These databases (there is a separate one for each disease process) automatically calculate the outcome parameters described below, where appropriate, permitting continuous evaluation of our RGS program.

To adequately evaluate the success of a RGS program, it is best to establish a common vocabulary that describes the outcome parameters of interest. These parameters can then also be used to critically appraise the relevant literature, providing a common basis for comparison between published results and local results. My evaluation of the SLN literature has led to my use of the following terms to describe SLN outcomes: **success rate, yield, positivity, sensitivity, false negative rate, exclusivity, and local failure rate.**

Success Rate: The ability to locate SLN in all basins identified by preoperative lymphoscintigraphy. Note that in the literature, this is not the biopsy of all nodes identified on nuclear scans; it is the biopsy of at least one node from each basin found to contain at least one SLN by lymphoscintigraphy. The expected rate for melanoma should approach 100%. The rate for breast cancer, with its documented longer learning curve, should be above 90% within the first 20-30 cases.

Yield: Number of sentinel nodes biopsied per patient (sometimes broken down per basin). Generally, for both melanoma and breast cancer, a mean of 1.5-2.0 SLN are biopsied per patient, with an upper range of 5-6 nodes.

Positivity: Percentage of patients who will have lymph node metastases identified by SLN pathologic examination. This number varies with patient diagnosis and known primary lesion prognostic factors, principally tumor size for breast cancer and Breslow thickness and ulceration for melanoma. This rate will also be affected by the type of pathologic evaluation performed, as discussed earlier in this chapter and in other chapters as noted.

Sensitivity/False Negative Rate: The ability of SLN biopsy to identify all patients with lymphatic metastases. These numbers are related by the equation:

$$100 = \text{Sensitivity} + \text{False Negative Rate}$$

As previously stated, the goal of SLN biopsy is to maximize the sensitivity while minimizing the false negative rate. This goal justifies the performance of the extended, comprehensive, pathologic evaluation algorithm. From a cancer prognosis and treatment perspective, probably the most important outcome variable is the false negative rate. If this variable result is unacceptably high (based on literature values, greater than 5%),[1] then patients in that treatment cohort will have too great a chance of developing regional or distant recurrence after being treated as "node-negative". Also, once you and your institution move beyond the routine performance of complete node dissection regardless of SLN pathology, sensitivity and false negative rate are impossible to calculate and can only be referenced historically. Therefore, as any institution begins to perform SLN, particularly for breast cancer where the standard of care remains complete axillary node dissection, these two variables must be calculated during the IRB protocol phase. The statistical endpoint of the IRB protocol that I have written for breast cancer, described earlier, is patient accrual high enough to document statistical equivalence of the cohort's false negative rate for SLN biopsy to published results, or 5%.

Exclusivity: This is the incidence of patients whose *only* nodal metastases are in SLN. The early publications of melanoma SLN biopsy all identified an interesting phenomenon, that patients with sentinel node metastases rarely had non-SLN metastases.[15] This is consistent with the concept of the SLN as *the primary draining lymph node* of the primary cancer site with the corollary that the other, nonsentinel lymph nodes in the basin had a significantly lesser role in "draining" that site of the body, and were therefore at substantially lower risk for metastatic involvement. More recent publications have placed the exclusivity rate of SLN in melanoma patients above 90%. Interestingly, the exclusivity rate for breast cancer patients is lower, at 65-70%.[16] The reasons for this difference is unclear, but may be due to intrinsic differences in the lymphatic drainage from skin versus breast tissue.

Local Failure Rate: Incidence of nodal metastases occurring after negative SLN biopsy of that lymphatic basin. This is one of the potential poor outcomes of a negative SLN biopsy, particularly one done without following the comprehensive pathologic evaluation schemes described in this handbook. A recent publication suggests that the local failure rate for melanoma patients following negative SLN biopsy is at least partially affected by the type of pathologic examination performed.[12] The local failure rate for breast cancer patients following negative SLN biopsy without routine completion axillary node dissection is completely unknown.

The outcome measurements for MIRP are substantially different. Since this is a newer procedure, employing a novel technical approach to a standard operation, I believe the outcome measurements currently should concentrate on localization parameters, criteria for parathyroid identification and success rates. Accordingly, we record information about the relative differences between cps measured over the diseased gland and elsewhere in the neck. We also record the type of instrumentation used and its settings. Finally, we record the cps ex vivo, operative time, and the pathologic and clinical outcome. This information is intended to facilitate future procedures through eliminating operative steps or frozen section analysis, or providing other technical "tricks" to make the procedure go more smoothly.

BILLING AND REIMBURSEMENT

This is simultaneously the most difficult, most challenging and most controversial and problematic section to write or edit in this handbook. It may also be perceived as the most valuable section by the general practitioner, who after all would like to reasonably and appropriately be compensated for the performance of a specialized operative procedure that requires special training and perhaps national credentialing. During talks that I give on RGS, discussions of billing and reimbursement are the most intently listened-to sections of my presentation.

It should come as no surprise to experienced surgeons that there is no existing CPT code that even comes close to describing the procedure performed during sentinel node biopsy.[17,18] The pathologists and nuclear medicine radiologists have all performed their portions of the SLN procedure before, even if not in the exact same fashion, and therefore billing for those segments of an SLN biopsy is straightforward. For MIRP operations, Sestamibi scanning and the pathologic examination of the parathyroid are standard procedures that again are covered by available CPT codes.

No discussion of billing and reimbursement in this era would be complete without a disclaimer: *Given the uncertain nature of billing for any new procedure, the following codes are given as examples and are based on (limited) experience and (limited) published advice from the AMA. These are based on our current thinking in clinical practice and are not intended to be suggestions to other physicians or office managers. It is possible that by the time this handbook is published, the situation may have changed completely and these coding examples may no longer be appropriate or viable. Any physicians who employ these codes would be doing so at their own risk and agree to hold the author, editors, and publisher blameless.*

Table 1.1. Billing for RGS procedures

Procedure	CPT Code	Description	Amount*
SLN Biopsy, Melanoma			
	38999	Unlisted lymphatic procedure	Varies**
	78195-26	Intraoperative lymphatic mapping	
SLN Biopsy, Breast CA			
On Phase I Protocol	38745	Axillary lymphadenectomy, complete	
	78195-26	Intraoperative lymphatic mapping	
After Protocol complete	38999	Unlisted lymphatic procedure	See axillary SLN**
	78195-26	Intraoperative lymphatic mapping	
MIRP			
	60500	Parathyroidectomy	
	78070-26	Intraoperative parathyroid mapping	

* Amount billed is usual and customary rate, except as otherwise noted
** For CPT code 38999, we bill the following amounts:

$1,000	SLN biopsy, inguinal node (up to two nodes)
$1,200	SLN biopsy, axillary node (up to two nodes)
$1,400	SLN biopsy, head and neck node (up to two nodes)
$100	Additional amount for each pair of SLN after first two in basin
75% of sum	SLN biopsy, more than one basin

(e.g., bilateral axillary SLN biopsy = (1,200 + 1,200) x 0.75 = $ 1,800)

We believe that in the absence of an established all-encompassing code for sentinel node biopsy, surgeons should code for the surgical removal of the SLN(s) and (separately) for the professional component of interpretation of the readout from the gamma detector device to localize the lymph nodes at risk. The second portion can be billed using an existing CPT code of 78195 with the modifier "-26". The CPT code 78195 is for nuclear medicine scanning to localize lymphatic drainage. We add the text, "intraoperative lymphatic mapping," to distinguish what we have done from the work of the radiologists who will likely also be using that same code, without the modifier, for the lymphoscintigraphy itself. The modifier is added to emphasize the interpretation aspect of the portion of the procedure we are billing for. The existing CPT codes for lymph node biopsy describe biopsy of nodes from a limited number of locations (CPT codes 38500-38530). For a few basins, they differ for superficial or deep nodal location, but in general describe only the act of biopsying an otherwise unspecified lymph node or nodes. We believe that these codes are inadequate to describe what occurs during SLN biopsy and do not appropriately reimburse the surgeon for the technically demanding procedure that required additional training and the purchase of a novel device. We therefore code all melanoma SLN procedures using 38999, unlisted lymphatic procedure. We enclose a two page explanatory letter with each bill submitted, describing what SLNs are, why they are biopsied, how they are biopsied, and what

Table 1.2. Reimbursement for RGS procedures*

Payor**	% Reimbursed***	Notes
Insurance Co #1	50%	
Insurance Co #2	100%	HMO
Insurance Co #3	15-25%	
Insurance Co #4	100%	HMO
Insurance Co #5	85%	
Insurance Co #6	100%	small private insurer
Insurance Co #7	60%	HMO
Insurance Co #8	Under appeal	"Not part of benefit package"
Insurance Co #9	Under appeal	"Included as part of wide excision"

* This experience is based on billing as described in Table 1.1, after about 50 melanoma cases in 1998
** Payors not identified
*** This is the percentage of the billed amount for CPT 38999 paid
    (All payors immediately reimburse for CPT 78195-26 at usual rates)

our fee schedule is for all SLN biopsies. We also enclose a copy of the operative report, with the portion describing sentinel node mapping and biopsy highlighted. This provides the payor with what we hope is adequate information about the procedure in general to enable them to reimburse the surgeon. Further, it shows them that we believe this to be a procedure that will be performed multiple times in the future, so that we have established a consistent fee schedule, accounting for different basins, their relative complexity from a surgical perspective, and the number of nodes biopsied. This fee schedule is shown in Table 1.1. Currently, we are performing all breast SLN procedures under IRB protocol with routine axillary node dissection. As such, we code only for the deep axillary node dissection (CPT 38745) for the procedure. In the future, we will likely adopt the same billing protocol for breast SLN as used for melanoma cases. We bill MIRP procedures using the standard parathyroidectomy code (CPT 60500).

The second portion of the bill for all SLN procedures is 78195-26. The complete axillary node dissections that we are currently performing after all breast SLN biopsies are also billed with this code, in addition to CPT 38745, as described above. The MIRP procedures have the corresponding code of 78070-26 for the intraoperative mapping of the parathyroid adenoma location.

The first test of a billing strategy is the reimbursement rate. To date, all insurance companies that we have billed have immediately paid on 78195-26. This amount is generally in the $50-$100 range. As might be expected the SLN biopsy billing using CPT 38999 is more problematic and has produced a variety of responses. The results of these policies five months after implementation are shown, without corporate or governmental identifiers, in Table 1.2. To date, this billing strategy seems appropriate and has been received relatively well by third party payers, but we anticipate that as SLN becomes more prevalent, the AMA will create new CPT codes for these procedures.

CONCLUSIONS

This chapter has described the implementation of a successful RGS program, from initial conception and organization, through outcome analysis and billing/reimbursement. The goal of this chapter was to "fill in the cracks" in any efforts to establish a sentinel node and radioguided surgical program. Any additional questions may be forwarded to me at e_whitman@hotmail.com.

ACKNOWLEDGMENTS
The author would like to acknowledge the assistance of Ms. Anita Boatman, RN and Ms. Patricia Eichholz, both indispensable to any smoothly running surgical practice.

REFERENCES

1. Krag D, Weaver D, Ashikaga T et al. The sentinel node in breast cancer: A multicenter validation study. N Engl J Med 1998; 339:941-946.
2. Reintgen D, Cruse CW, Wells K et al. The orderly progression of melanoma nodal metastases. Ann Surg 1994; 220:759-767.
3. Albertini JJ, Lyman GH, Cox C et al. Lymphatic mapping and sentinel node biopsy in the breast cancer patient. JAMA 1996; 276:1818-1822.
4. Veronesi U, Paganelli G, Galimberti V et al. Sentinel-node biopsy to avoid axillary dissection in breast cancer with clinically negative lymph-nodes. Lancet 1997; 349:1864-1867.
5. Giuliano AE, Jones RC, Brennan M, Statman R. Sentinel lymphadenectomy in breast cancer. J Clin Oncol 1997; 15:2345-2350.
6. Reintgen D, Haddad F, Pendas S et al. Lymphatic mapping and sentinel lymph node biopsy. In: Care of the Surgical Patient. Scientific American, 1998:1-17.
7. Emilia JCD, Lawrence Jr,W. Sentinel lymph node biopsy in malignant melanoma: the standard of care? J Surg Oncol 1997; 65:153-154.
8. Reintgen D, Balch CM, Kirkwood J, Ross MI. Recent advances in the care of the patient with malignant melanoma. Ann Surg 1997; 225:1-14.
9. Balch CM, Soong S, Bartolucci AA et al. Efficacy of an elective regional lymph node dissection of 1 to 4 mm think melanomas for patients 60 years of age and younger. Ann Surg 1996; 224(3):255-266.
10. Kirkwood JM, Strawderman MH, Ernstoff MS, Smith TJ, Borden EC, Blum RH. Interferon Alfa-2b Adjuvant therapy of high-risk resected cutaneous melanoma: The Eastern Cooperative Oncology Group Trial ECOG 1684. J Clin Oncol 1996; 14:7-17.
11. Norman J, Chheda H. Minimally invasive parathyroidectomy facilitated by intraoperative nuclear mapping. Surg 1997; 122:998-1004.
12. Gershenwald JE, Colome MI, Lee JE et al. Patterns of recurrence following a negative sentinel lymph node biopsy in 243 patients with stage I or II melanoma. J Clin Oncol 1998; 16:2253-2260.
13. Turner RR, Ollila DW, Krasne DL, Giuliano AE. Histopathologic validation of the sentinel lymph node hypothesis for breast carcinoma. Ann Surg 1997; 226:271-278.
14. Whitman ED, Frisse ME, Kahn MG. The impact of data sharing on data quality. Proceedings of the American Medical Informatics Association 1995; 19:952.(Abstract)
15. Joseph E, Brobeil A, Glass F et al. Results of complete lymph node dissection in 83 melanoma patients with positive sentinel nodes. Ann Surg Oncol 1998; 5:119-125.
16. McMasters KM, Giuliano AE, Ross MI et al. Sentinel-lymph-node biopsy for breast cancer - not yet the standard of care. N Engl J Med 1998; 339:990-994.
17. Coding consultation: Lymph nodes and lymphatic channels. CPT Assistant 1998; 8:10.
18. Physicians' current procedural terminology: CPT '98. Fourth edition, American Medical Association, Chicago, IL, 1997.

Intraoperative Gamma Radiation Detection and Radiation Safety

David A. Hillier, Henry D. Royal

INTRODUCTION

Effective intraoperative use of a gamma probe requires familiarity with basic principles of radiation, radiation detection and radiation safety. Basic radiation terminology, radiation detection, gamma probe testing, proper gamma probe use, radiation safety and regulatory issues will be presented in this chapter.

BASIC RADIATION TERMINOLOGY

The terms used to describe radiation are confusing and complex. To make matters worse, two different sets of terms are commonly used to describe radiation related quantities. The United States has persisted in its use of older terms whereas much of the rest of the world has adopted newer terms call Standard International (SI) units. Despite this difficulty, mastery of a basic radiation vocabulary is necessary in order to understand some important fundamental concepts regarding radiation and in order to effectively communicate with other medical personnel who routinely work with radiation.

Radiation is a general term describing the outward propagation of energy in any one of a wide variety of forms (Fig. 2.1). A broad distinction is made between "ionizing" radiation, which has sufficiently high energy to strip electrons from atoms, and the less energetic "nonionizing" radiation.

Ionizing radiation is subdivided into non-particulate and particulate radiation. Non-particulate ionizing radiation (photons) consists of x-rays and gamma rays. Higher energy x-rays and gamma rays penetrate the body, are detectable outside the body and therefore are useful for imaging. X-rays and gamma rays are, by definition, distinguished by the way in which they are formed and are

Radioguided Surgery, edited by Eric D. Whitman and Douglas Reintgen.
© 1999 Landes Bioscience

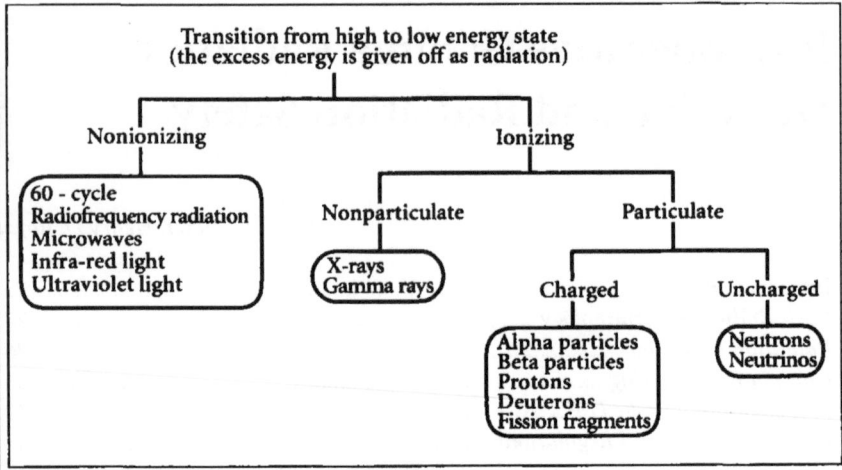

Fig. 2.1. Types of radiation.

indistinguishable once they have been emitted. X-rays result from transitions in the energy state of an electron. Gamma rays and the particulate forms of ionizing radiation are created when the nucleus undergoes a transition from a high energy state to a lower energy state. Gamma rays generally are of higher energy than x-rays, but there is considerable overlap. The magnitude of the energy content of a photon is expressed in electron volts (eV). Visible light photons (one form of nonionizing radiation) have energies of a few eV. X-rays and gamma rays used in diagnostic imaging have energies in the range of tens to hundreds of kilo-electron-volts (1000 eV = 1 kilo-electron volt or 1 keV).

Charged particulate radiation (e.g., alpha-rays and beta-rays) does not penetrate the body well and causes much greater biological harm than similar amounts of x-rays and gamma-rays. Charged particles can be used for therapy but not for imaging. Uncharged particles (e.g., neutrons and neutrinos) do penetrate the body, but are not easily detectable and neutrons have unfavorable dosimetry.

The words "radiation" and "radioactive" are often confused. An atom that is unstable spontaneously gives off radiation and is therefore radioactive. In contrast, an x-ray machine is not radioactive since it cannot spontaneously give off radiation (without an external power source). Patients who have had a chest radiograph do not spontaneously emit radiation. In contrast, patients who have been injected with a radioactive material will continuously emit radiation for a period of time.

The amount of radiation that patients emit from diagnostic nuclear medicine studies is quite small and not a significant problem (see the section on radiation safety below). The amount of radiation emitted decreases with time due to physical decay of the radionuclide (defined by a physical half-life) and elimination from the body (defined by a biological half-life). The physical half-life ($t_{1/2p}$) depends on the radionuclide (radioactive atom or radioisotope). The most commonly used radionuclide, Technetium-99m (Tc-99m), has a $t_{1/2p}$ of 6 hours. This means that the amount of radiation decreases by a factor of two every 6 hours from physical decay alone. The biological half-life ($t_{1/2b}$) depends upon the chemical compound to which the radionuclide is bound. The combination of the radionuclide and carrier molecule is termed a "radiopharmaceutical". In the United States, the most commonly used radiopharmaceutical for sentinel node lymphoscintigraphy is filtered Tc-99m sulfur colloid. For this radiopharmaceutical the rate at which the radiation decreases is primarily determined by physical decay since its $t_{1/2b}$ is very long.[1,2]

The amount (activity) of a radionuclide is expressed in terms of the number of atoms that are decaying (disintegrations) per unit time (second). This term is used because the number of counts detected by a radiation meter is generally directly related to the amount of radionuclide. For reasons discussed below, it is difficult to simply relate the amount of a radionuclide to the dose that a worker or patient might receive. In the United States, the unit for the measurement of the amount of activity is the curie (3.7×10^{10} disintegrations/second). The curie approximately equals the number of atoms disintegrating per second in a gram of radium. The SI unit for the amount (activity) of a radionuclide is the becquerel (1.0 disintegration/second). Curie amounts of radionuclides are not used in diagnostic studies. For most diagnostic studies, millicurie (mCi; 1000 mCi = 1 curie) amounts of a radionuclide are injected. For lymphoscintigraphy, microcurie (μCi; 1000 μCi = 1 mCi = 37 megabecquerels [MBq]) amounts are transported to the lymph nodes.

As a first approximation, the potential harmful effects of radiation are related to the amount of energy that was deposited from the radiation per gram of tissue. The unit used for absorbed dose is the rad (1 erg/ gram of tissue). The corresponding SI unit is the gray (1 gray = 100 rads). Another important determinant of the possible effects of radiation is the type of radiation. Some types of radiation are more damaging than others for the same amount of energy deposited because the energy is deposited more densely. For example, alpha rays can cause up to 20 times as much damage as gamma rays. The unit that takes into account the biological effectiveness of the radiation is called the rem. The SI unit for the rem is the sievert (1 sievert [Sv] = 100 rems). One rad of alpha rays might equal 20 rems but one rad of gamma rays equals one rem. A rem also equals 1000 millirems (mrem). Other determinants of the possible effects of radiation include the portion of the body exposed and the period of time over which the exposure took place.

The average radiation exposure to members of the public from natural sources of radiation (radionuclide in the soil and air, our food and bodies) is about 300 mrem per year. The occupational exposure limit is 5,000 rems per year to the body. The radiation dose to patients from lymphoscintigraphy would be less than 100 mrem.

The dose to workers from radionuclides is related to the amount of activity, the type of radiation emitted, the distance and the time spent near the source of radiation. Lead aprons are not as effective at decreasing the dose from most radionuclides as they are from x-rays because of the relatively higher photon energy of the gamma rays. Approximately 75% of the gamma rays from Tc-99m will be stopped by a standard 0.5 mm-thick lead apron compared to 95% of diagnostic x-rays.[3] The dose to patients is also related to the amount of activity and the type of radiation emitted, however patients have little control over the distance and the time spent near the source of radiation. The physical and biological half lives and the biodistribution of the radiopharmaceutical determine how long the source of radiation is in the patient.

RADIATION DETECTION

A variety of instruments can be used to detect gamma rays, including gas detectors (ionization chambers, proportional counters and Geiger-Mueller counters), organic liquid scintillation, x-ray film and solid detectors. Intraoperative gamma probes utilize solid detectors which efficiently absorb gamma rays due to high mass density and high atomic number.

Two basic types of intraoperative gamma probes are in current use. Sodium iodide crystals have been used for many years to detect radiation and are used extensively in current day gamma cameras. When radiation interacts with a sodium iodide crystal, the energy of the gamma ray is converted into a flash of light called a scintillation. This flash of light is converted to an electrical pulse of current by a photomultiplier tube that is optically coupled to the crystal. The intensity of the scintillation and therefore the size of the current pulse is proportional to the energy of the photon that was detected. This type of detector is called a scintillation detector. Approximately 85% of 140 keV photons from Tc-99m will be absorbed by a 3/8-inch-thick (9.5 mm) NaI crystal. Only a few of the absorbed photons reach and are detected by the photomultiplier tube, such that approximately 3 light photons are detected for each keV of energy absorbed.[1]

A newer type of solid detector uses a semiconductor to directly convert the energy of the detected photons to an electrical pulse of current. The semiconductor used in currently available intraoperative probes is cadmium telluride (CdTe). Approximately 60% of 140 keV photons from Tc-99m will be absorbed by 2-mm-thick CdTe. Charge carriers are produced for each 4.4 eV of energy absorbed. The energy resolution of semiconductor detectors is better than with scintillators since many more charge carriers are created for each gamma ray absorbed than are light photons detected for each gamma ray absorbed in a scintillation detector.[1,4-6]

GAMMA PROBE TESTING

Prior to using a new gamma probe in the operating room, a number of tests should be performed. This testing serves two purposes. First, testing ensures that the probe is functioning properly. Second, testing helps the surgeon become familiar with the use of the probe. Tests should be performed to determine the directionality (isocount lines), shielding of the probe and count rate capability (linearity).

Directionality can be determined by recording the count rate detected by the probe as a radioactive source is moved in precise locations around the probe. These measurements define the volume of tissue in which activity is detected by the probe. In the simplest of terms, this volume of tissue can be visualized as a cone with the apex directed towards the probe (see Fig. 2.2). The count rate detected from the same amount of activity will be greatest in the center of each slice through the cone and in the slices closest to the probe. In vivo, the count rate will dramatically decrease with depth due to the inverse square law and to attenuation of activity by the overlying tissue. The size of the cone of tissue sampled (field of view) by the probe can be readily modified by use of a collimator. A collimator is a metallic tube that limits the direction that gamma rays must come from in order to strike the detector. Most collimators have been optimized for use with Tc-99m. If radionuclides with higher energy photons (e.g., In^{111}-173 and 245 keV; I^{13}-364 keV; F^{18}-511 keV) are used, the collimator wall thickness is increased, shielding the detector from higher energy photons. The diameter of the collimator is usually similar to the diameter of the detector. A smaller field of view can be created if the detector is moved back from the tip of the collimator tube, or if the tube's length is increased relative to its diameter (see Fig. 2.2). Small fields of view offer the potential advantage of improved spatial localization and improved signal-to-noise ratio, but require a more careful search pattern to avoid missing the radioactive tissue. In addition, if the diameter is smaller, fewer counts will be detected. If

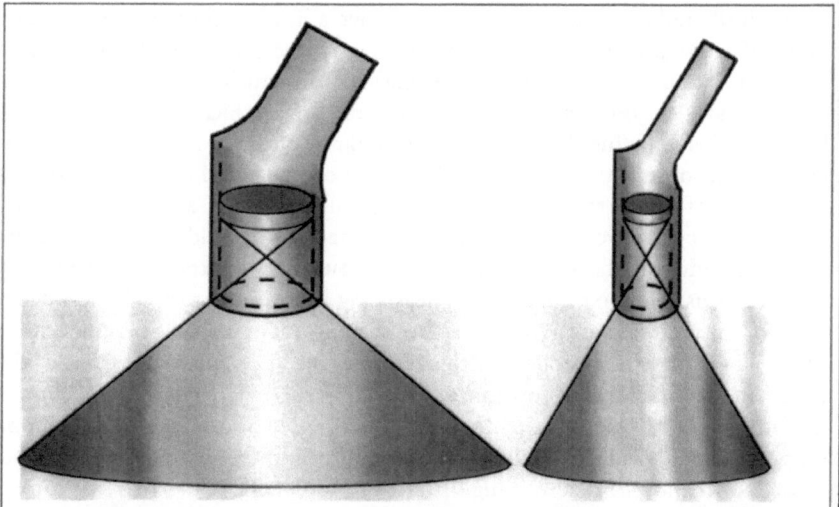

Fig. 2.2. Volume of detection for intraoperative gamma probes is a cone-shaped region. The volume is broad for a short collimator (left) and narrow for a long collimator (right). The latter offers the potential advantage of more precise localization, but disadvantages include increased probability of a sampling error and, under some circumstances, a lower count rate.

the diameter is unchanged, but the collimator is elongated, fewer counts may still be detected if the target does not fit inside the smaller cone of detection. How to optimize the tradeoff between maximizing localizing ability and minimizing sampling error is not clear and is dependent upon the technique utilized by the individual.

Another important characteristic of probe design that can be simply tested is the adequacy of side shielding. With lymphoscintigraphy, most of the injected activity remains at the site of injection. If the injection site is close to the site of the sentinel nodes, detection of the sentinel nodes could be compromised if there is inadequate side shielding. The adequacy of side shielding can be simply tested by measuring the count rate when a source of activity is in the center of the field of view compared to when the source is adjacent to the side of the detector. For the probes that we tested, the count rate from a Tc-99m source at the side of the detector was about 1% of the count rate when the source was in the center of the field of view. If the activity at the injection site is 100 times greater than the activity in the lymph node (a likely possibility) then equal numbers of counts may be detected from the injection site and the lymph node. Additional side shielding can readily be obtained when necessary by placing a small piece of lead (or a bulky metallic surgical instrument) between the side of the probe and the injection site. Additional side-shielding will particularly be necessary if the probe is used with radionuclides that have higher energy gamma rays. Based on tests performed on current probes, optimized for imaging with Tc-99m, the count rate from an I-131 source at the side of the detector can be as high as 50% of the count rate when the source was in the center of the field of view.

Typically radiation detectors will only have a linear response over a limited range of activity. When too much activity is used, the count rate that is detected reaches a plateau (non-paralyzable electronics) or can actually decrease (paralyzable electronics). Determining the range of count rates over which the probe responds linearly helps to define the useful operating range of the probe. The linearity of detector response with respect to activity level can be tested using two different methods. For either method, a small vial is filled with an amount of activity (1-10 mCi Tc-99m would be a good choice) that likely would greatly exceed the maximum amount of activity that the probe would ever be expected to detect in a discrete focus. The first method for measuring linearity uses the physical decay ($t_{1/2p}$) of the radionuclide in order to place varying but exactly known amounts of activity in front of the probe. For this method, the vial is placed in front of the probe, in exactly the same position, twice a day for several days. The date, time of day and count rate detected by the probe is recorded and plotted on a semilog graph. The graph will be a straight line over the linear response range of the probe. A second method of determining linearity uses metallic plates (e.g., 10 copper plates, 1/16 inch (1.6 mm) thick and approximately 3 inches (8 cm) square) to precisely vary the amount of activity that reaches the probe. This latter method is more convenient because it permits this test to be made more quickly. The copper plates are placed directly between the source and probe, one at a time and the count rate recorded. Since each copper plate absorbs a constant fraction of the activity, a semilog graph of count rate vs. the number of copper plates should be linear. The test can be repeated with a variety of probe-to-source distances in order to extend the count rate range over which the probe is tested. All probes tested demonstrated a linear response over a range of activity levels likely to be encountered during surgery.

The energy of a gamma ray can be partially absorbed in the body by a process termed Compton scatter (Fig. 2.3). The resulting gamma ray is of reduced energy and altered direction. It may therefore provide unreliable or confusing localizing information intraoperatively. Ideally this can be compensated for by manual or factory-set adjustments of the energy detection threshold of the gamma probe device, eliminating detection of the lower gamma radiation from Compton scatter.

PROPER GAMMA PROBE USE

If images are obtained prior to surgery, marking the location of the sentinel node on the patient's skin may help the surgeon locate the node. Marking the

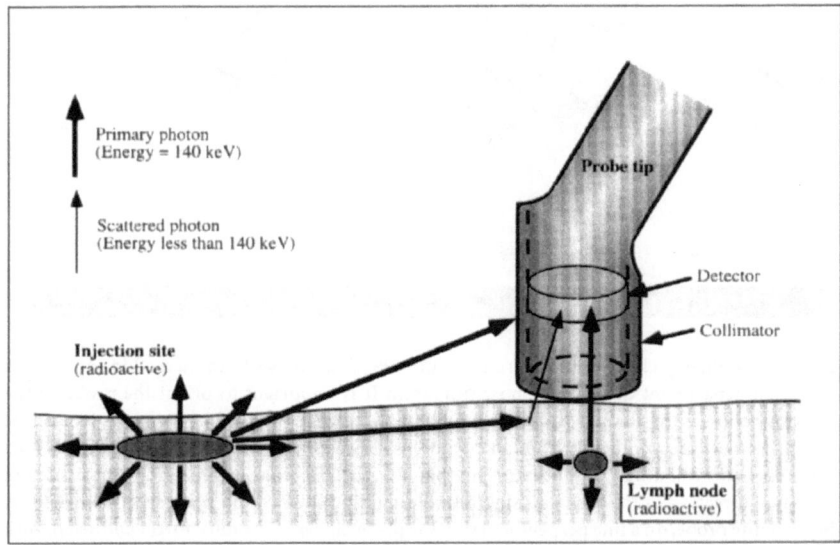

Fig. 2.3. Compton scatter. The injection site in lymphoscintigraphy contains a high level of activity relative to the lymph node. To localize the lymph node, the probe must have a collimator to block gamma rays from entering the side. Compton scatter, resulting from interaction of a primary photon from the injection site with an atom in the patient's tissue, may produce a photon that is directed into the probe detector, making detection of the lymph node difficult. A Compton scattered photon has a lower energy than the primary photon, and can be distinguished on this basis. The energy of the scattered photon is related to the scatter angle (the energy is nearly equal to that of the primary photon at low scatter angles). Energy resolution of the detector is therefore an important factor in scatter rejection and target-to-background ratio.

location of the node on the skin is subject to considerable error unless the following steps are taken. First, the skin should be marked with the patient positioned as he or she will be in during surgery in order to maintain the relationship between the relatively mobile skin and the underlying structures. Second, because of the problem of parallax, the skin mark site is dependent on the position of the gamma camera (see Fig. 2.4). Therefore, the person marking the skin must tell the surgeon the position of the gamma camera relative to the patient when the skin was marked.

Skin marking is performed by holding a container with a small amount of a radionuclide. The radionuclide is placed between the patient and gamma camera while watching a monitor that displays the node and the external source of activ-

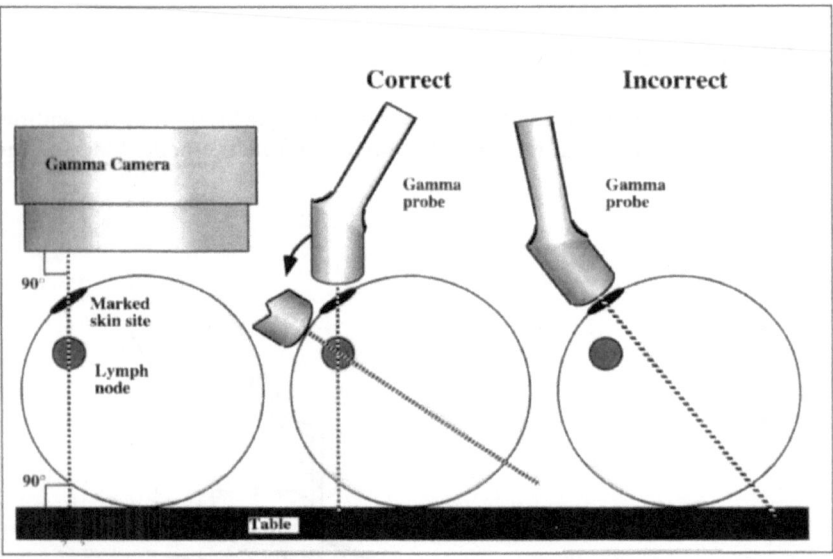

Fig. 2.4. Localizing a lesion intraoperatively after the skin site has been marked in the nuclear medicine department during lymphoscintigraphy. It is important to orient the probe in the same direction as the gamma camera to first localize the lymph node. The imaging plane of the gamma camera is commonly (although not always) oriented parallel to the floor. On a rounded body part, this will usually not be parallel to the skin surface. If the gamma probe is oriented in the same direction as the gamma camera (middle figure), the lymph node should be directly beneath the probe, on a line perpendicular to the probe tip face. The most direct approach, with the shortest distance from the skin surface, can then be determined by moving the probe away from the skin mark, perpendicular to the skin surface until the maximum count rate is obtained. The lymph node should be located at the intersection of the first line obtained with the probe at the skin mark and the line obtained with the probe perpendicular to the skin surface at the maximum count rate. If the gamma probe is oriented perpendicular to the skin surface initially, the lymph node will not appear directly beneath the skin mark, as shown in the diagram on the right. It is also important to take into account body positioning differences between that used during scintigraphy and intraoperatively. For example, if the neck is extended during surgery for a parathyroid adenoma, then the neck should also be extended during scintigraphy if the skin site is to be marked.

ity in real time. When the location of the external source corresponds to the sentinel node, the skin site is marked. Because these images consist of counts acquired over only a brief period of time, the quality of the images are poor and it may not be possible to locate some nodes which contain little activity.

Additional localizing information, may be obtained by marking anatomic landmarks with a radioactive marker or by using a large uniform source of activity (flood source). Most nuclear medicine facilities have a solid flood source that is used daily to check that the nuclear medicine camera is functioning properly. When a patient is placed between the uniform source of activity and the gamma camera, a transmission image is obtained that outlines the body (see Fig. 2.5).

Some surgeons use only the probe to find the sentinel node. The probe only approach may work in many patients and has the advantage of scheduling simplicity and reduced cost. However, there are a number of disadvantages in not obtaining pre-operative imaging. First, the lymphatic drainage of some lesions (e.g., a midline melanoma) may be uncertain. Second, in transit nodes may be missed in the case of an extremity melanoma. Third, sentinel nodes are not always identified with radioactive tracers. If nodes are not seen on preoperative images, dye based methods may be more useful in identifying the node intraoperatively.

The sentinel node is identified with a gamma probe by the fact that it has a higher amount of the radioactive tracer per gram of tissue (target) than does the surrounding tissue (background). For lymphoscintigraphy, the target/background ratio of activity is very high and therefore the sentinel node can usually be detected without much difficulty. Problems occur when the probe is not pointed in

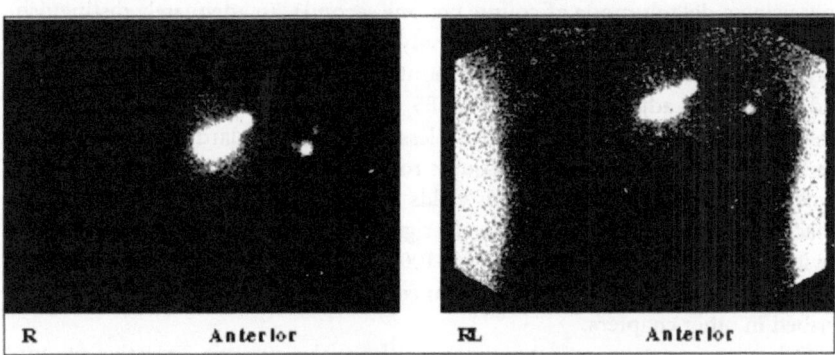

R   Anterior   RL   Anterior

Fig. 2.5. Transmission image. Lymphoscintigraphy examination in a 42-year-old woman with an anterior thoracic wall melanoma. Left: Anterior view, emission image. The gamma ray photons "emitted" from the radiopharmaceutical are imaged by the gamma camera to create an "emission" image. The intense activity at the injection site is in the center and uptake in three lymph nodes is seen to the patient's left. There is little anatomic information in such an image. Right: Anterior view, combined emission and transmission image. A Cobalt-57 sheet source was placed behind the patient during image acquisition. The patient's body blocks the majority of gamma rays from the sheet source (any that pass through the patient are described as having been "transmitted" through the patient) The unobstructed portion of the sheet source beyond the edge of the patient's body are transmits freely to the gamma camera, outlining the patient's body. One can clearly see that the uptake is in the patient's left axilla.

the right direction and is therefore not sampling the correct volume of tissue or includes some counts from the injection site. Additionally, the injection site contains much more activity than the node itself. If the node is near the injection site, additional shielding to block counts from the injection site may be helpful.

The count rate that is detected by the probe is greatly affected by the distance between the probe and the node. To the extent that the node is a point source relative to the probe, the count rate will go down as a function of $1/\text{distance}^2$. In addition, the tissue between the probe and the nodes absorbs some of the gamma rays. The net effect of these two factors is that the count rate detected from the probe greatly increases as the probe moves closer to the node. If the node contains very small amounts of activity, it may be difficult to confirm that it is a sentinel node in situ; however, once it is removed from the body, it can be counted outside the body and the count rate can be compared to that from a similarly sized piece of non-sentinel node tissue.

Currently available probes convey the count rate on a display (either an analog rate meter or digital display) and by sound. Sound provides immediate feedback and is most useful when easily detectable amounts of activity are present. Use of quantitative timed counts may be helpful in locating areas with smaller amounts of activity or poor target to background ratios.

From a purely hypothetical perspective, statistically different counts should vary by an amount greater than that expected simply due to chance alone, or at least three times the standard deviation. For radioactivity (defined by a Poisson statistical process), one standard deviation is the square root of the total counts. A common error is to use a time counting period that is too short (for example, by only using a default mode of counts per one second). To adequately distinguish the sentinel node from background activity, counting for a longer period of time may be necessary. For example, if the counts over region "A" is 100 in one second and that over an adjacent region, "B," is 85, there is no statistically significant difference. In this example, the difference is less than three standard deviations, where the standard deviation is 10 (the square root of 100) and three standard deviations is 30. If counts for 10 seconds yields 1000 over region "A" and 850 over an adjacent region, "B," this is a statistically significant difference (since the standard deviation is 33.3 and three times this is 100).[7] Other authors have used more operational definitions of the difference in counts defining a sentinel node, as described in other chapters.

Other common errors include moving the probe over the operative site too rapidly and varying the distance between the probe and the patient. Both of these errors will cause variations in count rate that are due to technical factors and therefore the count rate changes will not provide good localizing information. The easiest way to ensure a consistent distance is to hold the probe against the skin. When a lymph node is found but not yet resected, it may be lifted partially out of the surgical bed such that the probe can be angled toward the node and away from the injection site and other lymph nodes more effectively.

RADIATION SAFETY AND REGULATIONS

Radiation is frightening to many people. The reality is much less sensational. The potential harm from radiation exposure is proportional to the dose. The harm of greatest concern from occupational exposure to radiation is that the radiation will cause cancer. Although it has been possible to measure small (8%) increases in cancers in large populations instantaneously exposed to large doses of radiation (e.g., atomic bomb survivors), it has been difficult to measure any increase in cancers in occupationally exposed persons.

Since the harm from radiation exposure is proportional to dose, it is helpful to have some examples of the possible magnitudes of radiation doses. A useful benchmark is the amount of radiation we are all exposed to every year due to naturally occurring radioactive materials in our body and in our environment. Natural background radiation exposes every person in the U.S. to about 300 mrem per year. There is considerable variation in the amount of background radiation (± 100 mrem) from place to place and people rarely consider exposure to natural background radiation when choosing a place to live. There are some places in the world where normal background radiation exceeds the 300 mrem/year average by 1000s of mrem.

The U.S. Nuclear Regulatory Commission annual occupational dose limits for a radiation worker are 5 rem effective dose equivalent to the body (deep dose component of body badge dose), 15 rem to the eye, or 50 rem to skin or an individual organ. Occupational dose limits to pregnant women (who have declared the pregnancy) is 0.5 rem during the pregnancy. Typical whole body doses to patients who are injected for nuclear medicine studies is a few hundred mrem to a few thousand mrem. The average dose to nuclear medicine technologists is approximately 180 mrem/year.[8] The maximum allowable radiation dose to a member of the general public (from man-made sources other than medical exams benefiting the subject) is 100 mrem/year.

Table 2.1 lists estimated whole-body effective dose equivalent and finger doses from sentinel node biopsy, minimally invasive radioguided parathyroidectomy (MIRP) and fluoroscopy (e.g., during central venous catheter insertion). Based on the estimated per case radiation dose, the number of procedures that could be performed before exceeding three different annual dose limits are listed. The first limit (fourth column in Table 2.1) is the number of cases that results in the maximum allowable dose to a member of the general public (deep dose– 100 mrem/year). If it is anticipated that personnel may approach or exceed this level in a year, formal training as a radiation worker is necessary. This entails training in basic radiation safety concepts and may require a short written exam, administered by the institution's radiation safety office. Personnel are only required to wear a radiation badge if their annual whole-body radiation dose is likely to exceed 500 mrem, or if the finger dose exceeds 5 rem (10% of the annual occupational limit), as shown in the fifth column in Table 2.1. In practice, however, institutions may choose to issue radiation badges at a lower threshold (e.g., 100 mrem/year).

Note that the estimates in Table 2.1 are based upon worst-case calculations (see footnote for Table 2.1). The actual dose received is dependent on the conditions specific to each case, such as the activity administered, the time since injection, duration of the procedure, length of time spent near the source and the distance from the source. Based on the administered activity alone (20 mCi for minimally invasive radioguided parathyroidectomy [MIRP] and 500 mCi for sentinel lymph node biopsy), one can predict that the whole body radiation dose from MIRP would be substantially higher than the whole body dose from sentinel node biopsies. Personnel are only required to wear a radiation badge if their annual whole-body radiation dose is likely to exceed 500 mrem (collar badge), or if the finger dose exceeds 5 rem (finger dose). These thresholds represent 10% of the annual occupational limit, as shown in the fifth column in Table 2.1. However, since the surgeon and his staff likely have many sources of radiation exposure, a more conservative approach is suggested. If the annual number of procedures performed approaches or exceeds the number listed for the limit for a member of the public (100 mrem/year), monitoring is recommended. Although it is likely that the measured doses will be considerably less than the per case doses listed in Table 2.1, the most practical way to measure radiation doses accurately is with radiation badges. Because of possible radiation exposures from other sources, the per case dose can only be measured accurately if the radiation badge is only worn during the procedures of interest.

Good radiation safety practice encourages workers to take simple, easy to implement steps to make even small radiation doses smaller. This philosophy is referred to as the ALARA (an acronym for "As Low As Reasonably Achievable") principle. Unfortunately what is reasonable and what is simple and easy to do are very ambiguous and hence subject to considerable local variations. Local decisions should be based on facts and the benefits (small decreases in radiation dose) of invoking the ALARA principle must be weighed against the harms (e.g., delays in diagnosis, diversion of limited resources, distraction of the surgical staff).

There are three ways (i.e., time, distance and shielding) to significantly decrease radiation doses. The radiation dose is linearly related to time. Halving the time spent close to the patient would halve the radiation dose. The radiation dose is related to the square of the distance. Doubling the distance to the source will decrease the radiation dose by a factor of 4. Finally, placing a shield between the patient and operating room personnel can significantly decrease the radiation dose to the personnel. For lymphoscintigraphy, the injection site can be easily shielded by covering the injection site with a small thick (1-2 mm) lead sheet when access to the primary site is not needed. For MIRP, the radiopharmaceutical is much more widely distributed and more difficult to shield. Key operating room personnel could wear lead aprons (containing shielding of 0.5 mm thick lead equivalent) that would absorb 75% of the radiation; however, aprons are cumbersome and therefore impractical.[3]

Table 2.1. Estimated dose during procedure and regulatory limits for surgical and pathology personnel

Exposure	Personnel	Per case dose (mrem)	Limit to member of general public	Number of procedures per year	
				Radiation worker; level requiring monitoring	Radiation worker; maximum allowable
Sentinel node biopsy (following lymphoscintigraphy) (500 µCi Tc-99m)	Surgeon, body	0.29	350	1,700	17,000
	Surgeon, finger	6.6	–	750	7,500
	Pathologist, body	0.052	1,900	9,600	96,000
	Pathologist, finger	4.4	–	1,100	11,000
Minimally invasive radioguided parathyroidectomy (MIRP) (20 µCi Tc-99m)	Surgeon, body	2.1	48	240	2,400
	Surgeon, finger	10	–	500	5,000
	Pathologist, body	0.00094	110,000	530,000	5,300,000
	Pathologist, finger	0.85	–	5,900	59,000
Fluoroscopy, 3 minutes	Body, no lead apron	3.6-14	7	35	350
	Body, with lead apron	0.18-0.72	140	690	6,900
	Body with lead apron and thyroid shield	0.090-0.36	280	1,400	14,000

Estimated whole-body effective dose equivalent and finger doses for a single procedure are based on generalized worst-case assumptions, including no biological clearance, no attenuation by overlying soft tissues, and high (5%) uptake by the sentinel lymph node, and a large (2 g) parathyroid adenoma with uptake 15 times the average uptake of the remainder of the body. Typical doses to personnel during fluoroscopy on an adult at a distance of 3 feet are shown for comparison. A lead apron of 0.5 mm lead equivalent, worn by the operator, absorbs approximately 99% of x-rays in the diagnostic energy range and will reduce the deep whole-body effective dose equivalent by approximately 95%. It will absorb only 75% of gamma rays from Tc-99m at 140-keV energy.[3] It is assumed that, during surgery following lymphoscintigraphy, the surgeon will work at the lymph node site between hours 2 and 3 after injection (at a distance of 30 cm from the lymph node site and 60 cm from the primary site) and that he will then work at the primary tumor site between hours 3 and 4 after injection (at a distance of 30 cm from the primary site and 60 cm from the lymph node site). Surgeon finger dose is based uon a point source at 5 cm for 30 minutes at the primary site and at 1 cm for 15 minutes at the lymph node site. Following parathyroid scintigraphy, it is assumed that surgery is performed at hours 2.5 to 3.5 after injection. Surgeon dose for parathyroid scintigraphy is based upon the dose rate from patients 10 minutes after injection of 20 µCi of a Tc-99m labeled bone scintigraphy raiopharmaceutical, at a distance of 50 cm for body dose and at a distance of 10 cm for finger dose.[9] The pathologist is assumed to handle samples for 15 minutes immediately following surgical removal. Pathologist dose is calculated as that from a point source at 30 cm for body dose, at 5 cm for finger dose from the tumor primary site sample and at 1 cm for sentinel lymph node and parathyroid adenoma.

Alternatively a lead apron could be used to shield the patient's abdomen and lower chest. Sestamibi is excreted by the hepatobiliary system into bowel and by the kidneys into bladder. Therefore most of the activity will be in the abdomen at the time of surgery. Use of a lead apron over the patient's abdomen and lower chest would significantly decrease (perhaps 50% or more) the radiation dose to the surgical staff. Whether this should be done is controversial. A very small per case reduction of radiation dose (possibly 2.1 mrem to 1.0 mrem) may be achieved; however, there also may be some very small increased risk to the patient from respiratory insufficiency (due to the weight of the apron) or hyperthermia. Local decisions vary considerably due to the difficulty in assessing very small competing risks. An overriding principle is that radiation safety measures should never increase the surgical risk to the patient by increasing operating room time or otherwise interfering with the operation.

All fissionable materials and radionuclides derived from fission products fall under the auspices of the United States Nuclear Regulatory Commission (US NRC) in many states. In the remaining states, referred to as "agreement states," the regulatory control has been transferred to the state. The US NRC licenses radiation safety committees at individual institutions to oversee radiation safety functions. The US NRC also regulates disposal facilities. Typically the regulations in agreement states are patterned after those of the NRC. The pertinent regulations are found in Title 10, Chapter 1, Code of Federal Regulations –Energy, parts 20 and 35 (Nuclear Regulatory Commission, Washington DC 20555).

A recurring radiation safety question is whether surgical pathology specimens need to be treated differently if they are radioactive. Unfortunately, the answer is complex, confusing and at times, seemingly irrational. Generally, patients' excreta, body fluids and tissue are exempt from regulation. That is, following an injection of Tc-99m sestamibi, a patient can use a public toilet and eliminate several millicuries of Tc-99m into the sewer system. This exemption has been made so that the public is not denied the medical benefits provided by radionuclides because of cumbersome, impractical regulations. In contrast to a patient, a nuclear medicine technologist cannot empty a syringe containing several mCi of Tc-99m into a toilet. Technologists and other radiation workers must hold radioactive waste until it decays enough so no more radioactivity can be detected. This practice is called "decay in storage" or "DIS". For Tc-99m, DIS is not particularly burdensome since Tc-99m has a half-life of only 6 hours. After 10 half lives (60 hours), only 1/1000th of the original activity remains. In most cases, this small residual activity can no longer be detected and the waste can be disposed of in the appropriate nonradioactive waste stream. Standard procedures for DIS require that written records are kept documenting that waste was surveyed before disposal.

The exception to the exemption of patient excreta, body fluids and tissues from regulation is when the material is specifically collected to be assayed for radioactivity. For example, urine which is collected specifically for radioassay can no longer be disposed of by flushing it down the toilet. Once it has been assayed, the urine must be held for DIS.

Similar rules apply for surgical pathology specimens. If a specimen is radioactive because the patient happened to have a nuclear medicine study shortly before surgery, the specimen is exempt from regulations. On the other hand, if the patient was injected with a radioactive tracer prior to surgery for the purpose of radioassay, the specimen is subject to regulation. The exact requirements governing the handling of the specimen vary depending on the local regulations and the local interpretation of the regulations.

In states regulated by the Nuclear Regulatory Commission (NRC), the legal labeling requirements are specified in NRC regulation 10 CFR part 20.1904, with activity levels requiring labeling listed in appendix C. For Tc-99m this level is 1.0 mCi, while for some radionuclides it is more restrictive (this is especially true for I-131, which must be labeled if activity exceeds 1.0 mCi). If this activity level is exceeded, the container must be labeled with a radioactivity trefoil symbol, the radionuclide, amount of activity, time and date of the measurement, and the initials of the person who made the measurement. The majority of surgical pathology specimens contain insufficient radioactivity to require labeling by law. Specimens likely requiring detailed labeling include breast tissue from the injection site obtained after lymphoscintigraphy (the specimen may contain 1.0 to 2.0 mCi of Tc-99m) Tissue from the injection site in patients with melanoma are unlikely to require detailed labeling since usually less than 1 mCi of Tc-99m is injected.

For specimens containing less than 1 mCi of Tc-99m, no labeling may be necessary if the procedures for processing the specimens are reviewed by the radiation safety committee and they conclude that labeling is not necessary. Labeling would not be necessary if anticipated doses to the pathology staff are small (< 100 mrem/year; almost certainly true) and if DIS is already implemented since specimens are not discarded for a minimum of 5 days after surgery.

As shown in Table 2.1, the dose to a pathologist from handling radioactive tissue samples is very low and should not prevent initial processing of samples and preliminary histological determinations. Nor is it necessary to take special precautions for more complete processing. However, in the interest of taking simple measures that can make small doses even smaller (ALARA principle), personnel at some institutions delay final processing of the sample for 24 hours. The activity will have decayed to one-sixteenth its original value (for Tc-99m). This should only be considered if such a delay will not adversely affect patient care.

Based on the amounts of activity used in lymphoscintigraphy, it is not legally necessary to survey operating rooms that are used for lymphoscintigraphy. Simple calculations show that any anticipated dose rate to personnel from blood in the operating room is well below that allowed to members of the general public. The cleaning that is done to protect personnel and patients from significant biological hazards is more than sufficient for protection against radiation hazards.

Overall, it is important to emphasize that radioactivity exposure in the hospital setting must be monitored and approved by the radiation safety committee at that institution, particularly when it involves new procedures such as sentinel lymph node biopsy and MIRP. The guidelines presented in this chapter cannot be adopted at individual institutions without the guidance of these monitoring bodies. However, clinical data and radiation dose estimates clearly suggest that these procedures may be done safely at all hospitals with experience in handling these substances.

References

1. Sorenson JA, Phelps ME. Physics in Nuclear Medicine, 2nd edition. Philadelphia: W.B. Saunders 1987.

2. Thrall, Ziessman. Nuclear Medicine, The Requisites. St. Louis: Mosby 1995.

3. Huda W, Boutcher S. Should nuclear medicine technologists wear lead aprons? J Nucl Med Tech 1989; 17:(1) 6-11.

4. Serreze HB, Entine G, Bell RO, Wald FV. Advances in CdTe gamma-ray detectors. IEEE transactions on Nuclear Science. 1974; NS-21(1) 404-407.

5. Siffert P, Cornet A, Stuck R. Cadmium telluride nuclear radiation detectors. IEEE transactions on Nuclear Science. 1975; NS-22(1):211-225.

6. TerPogossian MM, Phelps ME. Semiconductor detector systems. Semin Nucl Med 1973; 3:343-365.

7. Gulec SA, Moffat FL, Carroll RG, Krag DN. Gamma probe guided sentinel node biopsy in breast cancer. Quarterly J Nucl Med 1997; 41(3):251-261.

8. Huda W, Bews J, Gordon K, Sutherland JB, Sont WN, Ashmore JP. Doses and population irradiation factors for Canadian radiation technologists (1978 to 1988). Can Radiol Assoc J 1991; 42:(4)247-252.

9. Harding LK, Mostafa AB, Roden L, Williams N. Dose rates from patients having nuclear medicine investigations. Nucl Med Commun 1985; 6:191-194.

Training and Credentialing Physicians in Radioguided Surgery

Eric D. Whitman 3

INTRODUCTION

As with any new surgical procedure, there are many technical aspects of radioguided surgery that must be mastered. In addition to the surgical techniques, this handbook also describes the multidisciplinary aspects of sentinel node procedures, specifically the essential roles of nuclear medicine and pathology in facilitating the identification of the sentinel nodes, and the optimal examination of the nodes after excision, respectively. This chapter approaches the evolving clinical applications of intraoperative lymphatic mapping from a more pragmatic viewpoint: how physicians are trained and subsequently credentialed to perform radioguided surgery.

TRAINING

Since the first reports of sentinel node procedures in the early 1990s, continuing medical education (CME) courses have been given at various institutions to train physicians to safely and correctly perform sentinel node mapping and biopsy. These courses were initially given at some of the first centers to publish their results. More recently, courses have been held at multiple other institutions. The quality and content of these courses is relatively unregulated; while most are structured to legally provide CME credits, there are no national standards or educational goals for training in sentinel node biopsy.

A typical course generally covers the following topics:
- Surgical techniques
  - Melanoma (sentinel lymph node, i.e., SLN)
  - Breast cancer (SLN)

Radioguided Surgery, edited by Eric D. Whitman and Douglas Reintgen.
© 1999 Landes Bioscience

- Other skin cancers (SLN)
- Parathyroid (adenoma resection)
• Nuclear Medicine
• Pathologic examination
- Melanoma
- Breast cancer, including touch prep
• Conscious sedation
• Observation of live surgical cases
• Animal laboratory of sentinel node biopsy (optional)

Newer additions to courses include the utilization of telemedicine to project interactive live audio and video content directly from the operating room to a large, remote, auditorium. Many courses appear to lack extensive discussion of some issues that are particularly relevant to the practicing surgeon, such as how surgeons should be credentialled in these techniques, how an institution might actually organize and promote a radioguided surgery program, and how these procedures are appropriately coded for billing purposes.

Empirically, what should the ideal training program for radioguided surgery include? Note first that I use the word, "program," instead of "course." In current usage, a *course* is generally an instructional event with a defined time frame, held at a place remote to where the actual clinical practice will occur. Courses, from a practical standpoint, have to condense information to fit the large amount of necessary clinical data into the time span of the course. A *program* would account for and emphasize the ongoing nature of any learning situation, where training takes place over a period of time and, potentially, places. A *program* would include a *course* component, but extend the effectiveness of the course format by including different educational modalities and allowing for proctoring or "post-training" evaluation of trainees.

The ideal training *program* for radioguided surgery would instruct interested physicians in the most up-to-date clinical techniques and outcomes data regarding these procedures, as is currently the practice. However, the ideal program would also include sessions on organizational aspects, credentialing issues, reimbursement, and data collection, all topics of crucial importance to practicing clinicians but often glossed over by course instructors, generally physicians at large academic medical centers with readily available capabilities in these areas. Most of the physicians attending courses will need to consider these issues carefully when they return to their home institution, as the infrastructure necessary to process and/or organize these essential components may not be in place. Therefore, the ideal training program should provide the clinicians in attendance with information that will assist them in the *logistical* aspects of initiating a radioguided surgical clinical service. This may include sample protocol documents, consent forms, data collection sheets, database program files, press releases, and letters to referring physicians, in short providing the program's students with a *business plan* describing how a radioguided surgical program is initiated and maintained.

Lastly, the ideal training program should enable the faculty or appointed surrogates to personally supervise the first few radioguided surgical procedures performed by the student. Current courses make no provision for this; a physician is trained for a day or two, then returns to his or her institution without being "checked-out" on the new procedures or necessary skills (on humans), putting pressure on the physician to make a "leap of faith" that the procedure(s) will proceed appropriately in his or her hands, and also putting pressure on the home institution to judge if this physician should be permitted to put patients at risk during performance of the procedure after such a limited instructional experience (i.e., credentialing, see below). Fixing this problem may involve the use of portable telemedicine modules, to lessen the travel needs of the faculty. Alternatively, "certified" or previously trained physicians in radioguided surgery may be commissioned by the course directors to act as surrogate instructors, completing forms that are mailed back to the training program director. This *proctoring* may also involve continued education and student evaluation via world wide web sites, where a student may need to correctly evaluate a set of images or plan patient care based on hypothetical patient histories posted on a web site, then submit their responses to the program faculty. This internet-based "post-testing" might also include instructional audio or video files that could be downloaded for a certain amount of time after official course attendance, based on password access with a 90 day expiration, for example.

To address these concerns, a newer education program is currently being designed, involving many of the authors of chapters in this textbook. The goal of this program is to comprehensively prepare all attendees to return to their home institutions and immediately be able to implement a clinically superior, well-organized, radioguided surgery program. The program will utilize telemedicine hardware and software to create a "virtual" learning experience, where class attendees could be at any one of a number of geographically separate sites and have an identical educational experience. Tentative course structure includes:

DIDACTIC LEARNING

- Surgical techniques: SLN for melanoma, breast cancer, other skin cancers, and other malignancies with emerging applications
- Surgical techniques: minimally invasive parathyroidectomy
- Nuclear medicine
- Radiation safety
- Lymphoscintigraphy
- Pathology
- Melanoma
- Breast cancer
- Coding/Billing/Reimbursement
- Credentialing issues
- Implementation of a radioguided surgical program. a business plan

EXPERIENTIAL LEARNING

- Observation of live surgery via telemedicine
- Multiple sites
- Multiple procedures
- Multiple surgeons
- All interactive with all audiences in real-time
- Animal laboratory of sentinel node mapping (remains optional)

This course is available beginning January, 1999 and information can be obtained by calling 1-888-456-2840.

The importance of excellent training in these evolving techniques cannot be overemphasized, particularly since most of the patients that are candidates to undergo radioguided surgery have cancer. Inaccurate or inappropriate use or implementation of these procedures could lead to devastating results in 5-10 years, as patients suffer the long-term ill effects of inadequate or misguided treatment of their cancers. Responsibility for this issue falls to both the leading experts in the field and the clinicians who introduce these techniques and procedures into their clinical practice. The experts must continue their investigations to optimize the technical and procedural aspects of radioguided surgery. They must also continue to dedicate their time and energy towards providing superior instruction for other clinicians in these techniques and towards the design and standardization of educational programs and credentialing criteria. The clinicians in some ways shoulder the more arduous task: they ultimately will treat the majority of patients with these techniques, and therefore it is imperative that all physicians receive adequate instruction, both didactic and experiential, and also establish guidelines at their own institutions for the safe and correct way to employ this technology.

CREDENTIALING

Although physicians of multiple specialties may (and do) attend the existing CME courses, the majority of attendees seem to be surgeons, and it appears that the majority of credentialing issues will center on the actual performance of the surgical procedure, not the lymphoscintigraphy and pathologic examination. Lymphoscintigraphy is a well-described technique,[1] and exact protocols to perform the radionuclide scan for radioguided surgical procedures are available from multiple published sources. Similarly, the histopathologic techniques are also well described in the medical literature.[2] The technical aspects of performing the necessary pathologic examinations are utilized by pathologists for other indications (or can easily be obtained through an outside specialty laboratory) and it therefore does not seem necessary in most cases to train pathologists in these techniques. The most important aspects of developing a radioguided surgical program at an institution from both the radiologic and pathologic perspectives will be the implementation by both departments of a consistent, reproducible approach to all patients undergoing these procedures (see chapter 1).

The emergence of radioguided surgical techniques and procedures is analogous in many ways to the rapid adoption of laparoscopic surgery less than a decade ago. In both cases, new technology and creative surgical perspectives combined to create a set of revolutionary surgical procedures. With laparoscopic surgery, almost overnight most if not all general surgeons in the United States desired training, and wanted to begin performing minimally invasive surgical procedures. Many hospitals were ill-prepared for the inevitable credentialing issues: how to regulate the conditions under which surgeons would be allowed to utilize these new techniques.

Unfortunately, when surgeons began performing laparoscopic procedures without formal training or without firm hospital or governmental regulatory policies, the incidence of cholecystectomy-related complications, particularly bile duct injuries, increased. In response to these problems, most hospitals established minimal criteria to allow surgeons to perform laparoscopic surgery. In New York state, a law was passed defining the training and experience necessary for a surgeon to be allowed to perform laparoscopic surgery. Despite these precautions, the incidence of bile duct injury remains 2.5-4.0 x higher after laparoscopic cholecystectomy than in the open cholecystectomy "era", in a recent study.[3]

The issues for radioguided surgery are also somewhat more complex, because the surgery itself is probably less morbid, or with less morbid potential, than the situation with laparoscopic vs. open cholecystectomy. However, radioguided surgery is to date used primarily to stage malignancies, so that errors in technique when performing these procedures might have a delayed adverse outcome, when the patient develops an unexpected recurrence months or years later. In this situation, inaccurate or inappropriate use of radioguided surgical techniques might have altered or deferred adjuvant therapy, with resulting life-threatening and/or medicolegal implications.

The American College of Surgeons is taking a proactive position on other emerging surgical technologies, specifically stereotactic breast biopsy,[4] and it appears likely physician qualifications for radioguided surgery will also be proposed by that organization. In a recent issue of the Bulletin of the American College of Surgeons (ACS), the Board of Regents announced that they had approved a "process by which its Fellows and Associate Fellows could be verified for the use of emerging technologies."[5] The intent of this laudable process is to provide surgeons with documentation that they have the necessary training and experience to be allowed by local credentialing facilities to utilize the new technology in patient care. Although not specifically mentioned in the announcement, the rapid emergence of radioguided surgery for multiple clinical indications suggests that this will be one of the early areas to be addressed by the Committee on Emerging Surgical Technology and Education (CESTE), the group charged with developing verification programs and standards.

At present, there are no universally adopted credentialing criteria for surgeons to be permitted to perform radioguided surgery. At national meetings and in major publications,[6] the responsibility for credentialing surgeons has been transferred to the individual hospitals, yet as previously noted there are no consistent training guidelines. Pending the establishment of such criteria and guidelines, particularly through the efforts of CESTE and ACS, the provisional regulations used at my home institution, Missouri Baptist Medical Center of BJC Health System in St. Louis, Missouri serve as an excellent example of how a hospital might control usage of radioguided surgery.

Our criteria are based on experience, training, and proctoring. Initially, only surgeons with prior experience at other hospitals or who had taken a CME-accredited course were given "Sentinel Node Privileges." This requirement was made very clear to both the surgeons and the operating room staff: except for a limited list of surgeons, no other surgeons would be allowed to perform radioguided surgery unless they:

1) Took a CME-accredited course in radioguided surgery, or provide documentation of residency training experience in radioguided surgical techniques;

2) Or, they could be proctored for a minimum of five radioguided surgical procedures by an experienced surgeon (on a voluntary basis by a local experienced surgeon; in all cases to date, the author); or through the mentoring service offered at Moffitt Cancer Center. In this case, the proctor would then send a letter to the Chief Surgeon of the hospital indicating that the trainee surgeon in question had been proctored on a certain number of cases and was proficient to perform radioguided surgical procedures independently.

In this evolving policy, one partially unresolved issue is what would happen if a surgeon went and took a CME-accredited course now, and then returned to the hospital seeking radioguided surgical privileges. It is not clear if lack of supervised experiential training in these techniques should then require one or two cases of local proctoring by an experienced surgeon to ensure that the trainee surgeon in question is indeed ready for independent performance of radioguided surgery. Our operating room administrators and physician leaders are currently addressing that issue, with plans to publish detailed guidelines for credentialing at our hospital.

What has happened to date at our institution is similar to what I understand is happening elsewhere: the local surgeons are taking advantage of the experience and interest of a solitary motivated colleague who is willing and able to train them in the techniques during surgical procedures on the trainee's own patients. For example, if a surgeon without radioguided surgical privileges is referred a patient who may be eligible for sentinel node biopsy, that surgeon would schedule the patient for surgery, coordinating the operative schedule with my calendar to ensure my full availability. The nonprivileged surgeon explains to the patient that this is a new technique and that their surgeon will be enlisting the aid of another local surgeon who is more experienced in the technique. In this way, the doctor-patient relationship is maintained, and the radioguided surgery may still be performed. This method of obtaining privileges in radioguided surgery has been more attractive to my colleagues than the time, expense, and travel necessary for an official CME-accredited course. As the courses themselves (see earlier section) and ACS/CESTE verification criteria evolve, it may become necessary for any future surgeons seeking radioguided surgical privileges to enroll in more formal training courses.

SUMMARY

As with any new technology, proper training and documented credentialing standards are essential to the safe and clinically prudent implementation of radioguided surgical procedures by physicians at individual hospitals. The lessons learned during the advent of laparoscopic cholecystectomy, where the incidence and type of operative biliary complications changed from the "open" era, should be built upon to prevent similar occurrences from radioguided surgery, particularly since most of the applications developed to date involve the treatment of malignancies, where adverse outcomes may not become clinically apparent for months or years.

Currently, training programs provide varying levels of didactic and experiential learning activities. Supervision or oversight/review (i.e., proctoring) of newly trained physicians after the course is complete is nonexistent. At our institution, we have implemented guidelines that ensure that patients only receive radioguided surgical procedures by those surgeons with documented training or experience in these techniques, or under the direct, hands-on, supervision of another surgeon with the necessary experience.

Ultimately, it is my hope that a large governing body such as the American College of Surgeons will step forward and take steps to ensure patients, third party payors, and other physicians that a given surgeon has obtained adequate training to allow him or her to utilize radioguided surgical techniques in clinical practice. This desirable level of credentialing can only be possible with standardization of educational objectives and experience by existing or future training courses.

ACKNOWLEDGMENTS
The author wishes to thank Dr. Douglas Reintgen, Dr. Charles Cox and Dr. James Norman, all of H. Lee Moffitt Cancer Center and the University of South Florida, Gene Theslof, of Infomedix Communications, Inc., and Eric C. Miller, of United States Surgical Corporation, for their invaluable thoughts and ideas on the subject of training programs for radioguided surgery.

REFERENCES
1. Reintgen D, Cruse CW, Wells K et al. The orderly progression of melanoma nodal metastases. Ann Surg 1994; 220(6):759-767.
2. Turner RR, Ollila DW, Krasne DL, Giuliano AE. Histopathologic validation of the sentinel lymph node hypothesis for breast carcinoma. Ann Surg 1997; 226: 271-278.
3. Strasberg SM, Hertl M, Soper NJ. An analysis of the problem of biliary injury during laparoscopic cholecystectomy. J Am Coll Surg 1995; 180:101-125.

4.   Physician qualifications for stereotactic breast biopsy: A revised statement. Bull Amer Coll Surg 1998; 83 (5):30-33.
5.   Verification by the American College of Surgeons for the use of emerging technologies. Bull Amer Coll Surg 1998; 83 (5):34-35.
6.   Reintgen D. The credentialing of American surgery. Ann Surg Oncol 1997; 4:99-101.

3

Melanoma Lymphatic Mapping: Scientific Support for the Sentinel Lymph Node Concept and Biological Significance of the Sentinel Node

Merrick I. Ross, Jeffrey E. Gershenwald

INTRODUCTION

The lifetime incidence of melanoma will reach an all time high of 1 in 75 individuals by the turn of the century.[1] Fortunately, as a result of increased public awareness and screening programs, the vast majority of newly diagnosed patients will present with clinically localized (stage I and II) disease. While patients with thin melanomas are likely to be cured following appropriate surgical excision of the primary tumor alone, a significant percentage of individuals with tumors that are thicker or exhibit other unfavorable prognostic factors, harbor clinically undetectable regional lymph node metastases. Such disease may serve as a potential source or predictor of subsequent nodal failure and distant dissemination. The routine use of lymphadenectomy targeted to those patients possessing an increased risk of microscopic nodal disease—an approach popularized as elective lymph node dissection (ELND)—has not consistently been shown to improve survival.[2,3] Accordingly, the role for ELND in the initial surgical management of these patients has been controversial.[4] Historically, the motivation to perform ELND was strong, primarily because surgery was the most effective melanoma treatment modality, since no effective systemic adjuvant therapy had yet been identified.

The major obstacles which prevent widespread acceptance of ELND are the following: 1) conflicting results from clinical trials which assessed survival of patients who received ELND compared to patients who only underwent lymphadenectomy as treatment for clinically detectable regional lymph node failure

Radioguided Surgery, edited by Eric D. Whitman and Douglas Reintgen.
© 1999 Landes Bioscience

(therapeutic lymph node dissection),[2] 2) unnecessary morbidity and costs associated with formal lymphadenectomy in patients without evidence of microscopic nodal metastases, and 3) the common (but never proven) belief that lymphadenectomy in patients with undetectable metastases may alter the regional immune system and render the patient vulnerable to recurrences.[5] Despite these arguments, those promoting ELND remain concerned about the delay in treatment of lymphatic disease until it is clinically palpable, when the tumor burden and the number of lymph nodes involved is greater and overall prognosis is worse.[6-8]

Amidst this long-standing controversy has emerged a rational approach to the clinically negative regional lymph node basin known as "selective lymphadenectomy", using the techniques of lymphatic mapping and sentinel node biopsy (LM/SNB), initially introduced in the surgical literature by Don Morton.[9] This approach is based on the concepts that the first draining lymph node—"the sentinel node"—is the lymph node most likely to contain melanoma metastases, and that a minimally invasive surgical technique can consistently and accurately identify this lymph node. While the initial motivation for LM/SNB was to perform lymph node dissections only in patients with proven microscopic disease (early therapeutic dissection) and spare those patients without nodal metastases the risks and cost of unnecessary surgery, the recently reported improvement in survival of node positive patients treated with high-dose interferon (Intron A, Schering Incorporated, Kenilworth, NJ) in the adjuvant setting has promoted accurate lymph node staging as another and probably more important motivation.[10]

SCIENTIFIC SUPPORT FOR THE SENTINEL NODE CONCEPT

Lymphatic mapping, as it applies to melanoma, relies on the hypothesis that the dermal lymphatic drainage from specific cutaneous areas to the regional lymph node basin is an orderly and definable process. These lymphatic drainage patterns should mimic how melanoma cells spread within the lymphatic compartment such that the first lymph node(s) receiving lymphatic drainage are the most likely to contain metastatic disease. In theory, each lymph node within a formal basin may potentially function as a sentinel node, but for different and finite regions of the skin. In addition, in a patient whose melanoma arises from a cutaneous site with potential drainage to more than one lymph node basin (i.e., trunk), sentinel nodes exist in each basin to which the primary tumor drains.

To test this hypothesis, a reliable method of SLN identification had to be established. Pioneers of this technique performed preclinical animal studies which evaluated a variety of dyes that could be intradermally injected and transported through the lymphatic system to the regional lymph node basin, thereby providing a visualization of the sentinel node upon surgical exploration of the lymph node basin. Morton and colleagues demonstrated that two dyes were most effective: isosulfan blue (Lymphazurin) and Patent Blue V.[9] Initial clinical studies were performed in melanoma patients to determine the following: 1) sentinel node identification rates, and 2) the accuracy of the sentinel node in establishing the presence or absence of regional nodal metastases.

The first report published in 1992 evaluated 237 consecutive patients using intradermal injections of Lymphazurin around the intact primary melanoma or excisional biopsy site.[9] The authors demonstrated an 82% sentinel node identification rate and an average of 1.3 SLNs were removed per basin. Subsequent studies from the M. D. Anderson Cancer Center, Moffitt Cancer Center, and the Sydney Melanoma Unit in Australia reported similar findings.[11-13] Accuracy assessment was similarly accomplished at all of these centers through the use of synchronous ELND performed at the time of the sentinel node biopsy in order to ascertain the false negative rate. For these studies, a false-negative event was defined as the detection of microscopic disease in a non-sentinel node when the sentinel node from the same basin was histologically negative. Accordingly, the false-negative rate was then calculated as the number of false-negative events divided by the total number of patients with microscopic nodal disease. Collectively, these initial studies evaluated 472 patients, 86 of whom were found to have regional node metastases. A false negative rate of 5% and an overall accuracy of 95% was established. These data strongly support the sentinel lymph node concept (Table 4.1).

Additional evidence that regional lymph node metastasis is an orderly and non-random event is provided from studies performed at the M. D. Anderson Cancer Center. Investigators found that the sentinel node was the only node involved in 83 (79%) of the basins with at least one metastasis, while additional disease was identified in only 21% of the lymphadenectomy specimens.[14a] A subsequent analysis demonstrated that 68% of all sentinel nodes and only 1.8% of all non-sentinel nodes in 105 therapeutic lymphadenectomy specimens contained metastatic disease when at least one sentinel node was positive for melanoma (Table 4.2).

As a result of the tremendous interest generated from these studies, many centers subsequently adopted the concept of selective lymphadenectomy when managing newly diagnosed intermediate and high risk Stage I and II melanoma patients.[27] Prospective clinical programs were established to enroll patients with the following primary tumor characteristics: Breslow tumor thickness of at least 1 mm or Clark's level of invasion greater than III, or the presence of histologic ulceration.

Table 4.1. Sentinel node identification and accuracy of initial studies: dye injection alone

Study	Year	Patients N	SLN ID Rate N (%)	+SLN N	+ELND N(%)	False-Negative Rate N(%)
UCLA	1992	237	194 (82%)	38	40 (21%)	2 (5%)
MDACC/MCC*,	1993, 1994	117	103 (86%)	23	24 (20%)	1 (4%)
Sydney Melanoma Unit	1995	118	105 (89%)	20	22 (21%)	2 (%)
Total		472	402 (85%)	81	86 (21%) 5 (5.8%)	

*M.D. Anderson Cancer Center, Moffitt Cancer Center

Table 4.2. *Additional metastases in therapeutic lymph node dissection after at least 1 positive SLN identified*

Total No. positive lymph node basins	105
Basins with additional positive SLNs, N (%)	22 (20)
Total No. LSNs harvested	183
Positive SLNs, N (%)	125 (68)
No. non-SLNs harvested	1883
Positive non-SLNs, N (%)	34 (1.8)

Adapted from Gershenwald et al. Improved sentinel lymph node localization in primary melanoma patients with the use of radiolabeled colloid. Surgery 1998; (In press).

Patients with a histologically positive SLN underwent therapeutic lymphadenectomy, while those with a negative SLN were observed. This large clinical experience yielded significant improvements in SLN localization techniques, generated additional findings which supported the sentinel lymph node concept, and provided valuable insights into the biologic significance of the SLN.

TECHNICAL ADVANCES

Although initial sentinel node biopsy experiences using the blue dye only technique provided a promising beginning, it was clear that room for improvement existed since initial sentinel lymph node identification rates were only 80% to 85%.[9,11-13] Prior experience with preoperative cutaneous lymphosintography has also improved the technique of SLN localization. This approach has provided a reliable and accurate way to identify regional lymph node basins at risk for metastases in patients whose melanoma arose in anatomic regions of ambiguous lymphatic drainage.[14,15] Application of this technology to lymphatic mapping clearly delineates the lymphatic drainage patterns from primary tumor injection sites to the regional nodal basin(s) at risk.[16] Afferent lymphatic vessels are well-visualized and relative location and number of sentinel nodes within the regional nodal basin(s) are established. Importantly, these techniques also helped to identify the uncommon but clinically relevant sentinel nodes that are located outside of the formal regional nodal basin, termed "in-transit" sentinel nodes.[16] Furthermore, the presence, if any, of direct lymphatic drainage to the popliteal or epitrochlear regions can also be established for primary melanomas arising distal to the knee or elbow, respectively. These high-resolution scans provide preoperative road maps for the surgeon and aid in the intraoperative identification of the blue-stained sentinel node(s).

The intraoperative use of a handheld gamma probe, capable of detecting the accumulation of intradermally injected radiolabeled colloid within sentinel nodes, provided the next important technical advance.[14a,17,18] This technique provided the surgeon with a sensitive instrument with which to transcutaneously localize the sentinel node prior to the skin incision and with an adjunct to the visual cues required to localize the blue-stained sentinel node at the time of surgical exploration. Preclinical animal models of lymphatic drainage performed by Nathanson provided useful information concerning the kinetics of different colloid preparations.[19] Clinical studies from the Moffitt and M. D. Anderson Cancer Centers demonstrated that the ratios of radioactivity within the sentinel node compared to neighboring non-sentinel nodes actually increased with time, providing surgeons a window of opportunity during which to perform the SLN biopsy after injection of radioactive colloid in the nuclear medicine department.[18] These data demonstrated that the particles in sulfur colloid preparations were large enough to be actively phagocytized by macrophages within a lymph node. The majority of material reaching the nodal basin was concentrated within the first draining lymph node—the SLN—-with only minor passage of colloid to secondary echelon (non-sentinel) nodes. Improved SLN identification rates using combined modality techniques compared to blue dye alone have been obtained from the Netherlands[20] and the M. D. Anderson Cancer Center.[14a] In these reports, identification rates increased to 99% with the largest incremental improvement in the cervical (neck) and axillary regions (Table 4.3).

In a multicenter trial reported by Krag et al,[17] identification rates of greater than 95% were obtained when injections of radiolabeled colloid alone were primarily employed. Unfortunately, concomitant ELNDs were not performed in this or other SLN studies in which radiolabeled colloid injections were utilized. As a result, questions have emerged whether or not the SLN localized in this manner was definitively the first lymph node of drainage and thus an accurate indicator of regional nodal metastases. The 16% incidence of SLN involvement reported by Krag was similar to the 20% incidence previously demonstrated with blue dye injections, and alleviated some of the concerns.[17] A recent study from the M. D. Anderson Cancer Center using both localization techniques more convincingly supported the concept that the node identified with the radiolabeled colloid is the sentinel node: 1) in patients with microscopic nodal disease and more than one sentinel node identified, the metastases were present within the lymph node containing the highest radioactivity 92% of the time, and 2) co-localization of blue dye and radioactivity occurred in at least 93% of the sentinel lymph nodes identified.[14a]

Table 4.3. SLN identification rate according to basin mapped and SLN biopsy technique

Basin	All Patients N (%)	Dye Alone N (%)	Dye & Colloid N (%)	P-Value
Cervical	42/47 (89)	5/9 (56)	37/38 (97)	0.002
Axillary	413/445 (93)	142/169 (84)	217/276 (98)	< 0.00001
Inguinal	222/227 (98)	93/98 (95)	129/129 (100)	< 0.03
Total	766/719 (94)	240/276 (87)	437/443 (99)	< 0.00001

Adapted from Gershenwald et al. Improved sentinel lymph node localization in primary melanoma patients with the use of radiolabeled colloid. Surgery 1998 (In press).

Interestingly, the number of sentinel nodes identified per nodal basin was greater when both modalities were employed compared to blue dye alone (1.74 versus 1.31)(Table 4.4).[14a] The following explanation is offered to account for this finding. The first blue lymph node encountered within a lymph node basin may be only one of the sentinel nodes present in that basin, or may actually be a secondary echelon node that received blue dye via the efferent lymphatic channels of a sentinel node located in a deeper, or perhaps different region of the basin. Failure to detect the sentinel node(s) or additional sentinel nodes may result from use of the blue dye alone, since it is difficult to completely examine the entire basin if one relies solely on visualization of blue dye. In contrast, intraoperative use of the gamma probe provides a method of detection that is independent of and more sensitive than visual cues. Blue-stained sentinel nodes that are initially undetected or sentinel nodes which did not receive a sufficient amount of blue dye to be visualized because of technical problems are likely to be located with the gamma probe (Fig. 4.1).

BIOLOGIC SIGNIFICANCE OF THE SENTINEL NODE

Studies have demonstrated that the incidence of SLN metastases correlates directly with increasing tumor thickness (Table 4.5).[14a] This strong correlation is nearly identical to that of the incidence of microscopic nodal disease in ELND specimens when stratified by tumor thickness.[5] Other known primary tumor factors such as anatomic location, ulceration, and Clark's level of invasion, also predict sentinel node involvement.[14a] In a multivariate analysis, the two variables that independently predicted SLN involvement were tumor thickness and the presence of ulceration (Table 4.6). The finding that the most powerful primary tumor prognostic factors were also the best predictors of SLN metastases offers further evidence that SLN involvement is a biologically important and non-random event.

Further support for the SLN concept is provided from long-term follow-up of patients who have undergone a negative SLN biopsy and no further surgical intervention in the lymph node basin. In a cohort of almost 250 patients followed for over three years, only 3% of the patients have died of recurrent disease, 10% of the

Table 4.4. SLNs identified according to SLN biopsy technique

	All Patients	Dye Alone	Dye & Colloid	P-Value
Mean	1.59	1.31	1.74	< 0.00001
Median	1	1	2	–
Range	1–7	1–4	1–7	–

Adapted from Gershenwald et al. Improved sentinel lymph node localization in primary melanoma patients with the use of radiolabeled colloid. Surgery 1998 (In press).

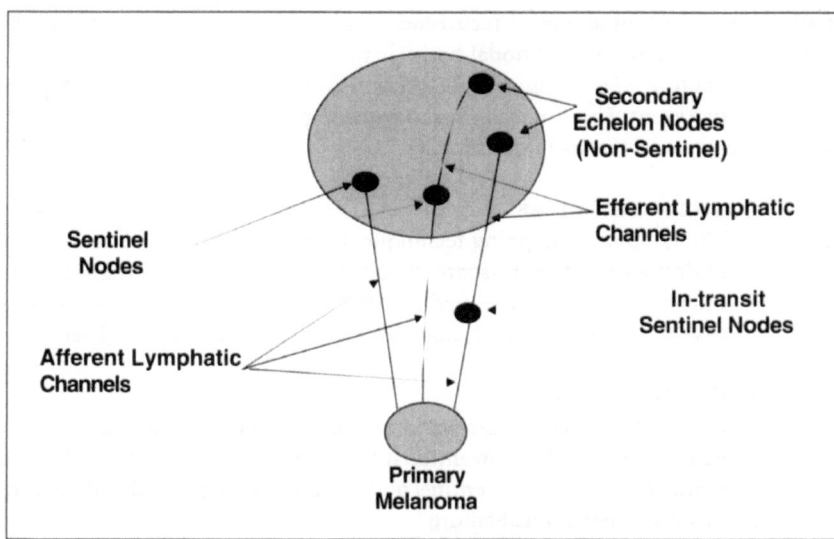

Fig. 4.1. Potential lymphatic drainage patterns. By definition, the first node(s) of drainage identified following injection of blue dye or radiolabeled colloid that travels through afferent lymphatics from the primary tumor are termed sentinel nodes. Some of the blue dye as well as small amounts of colloid may pass to secondary echelon nodes via efferent lymphatic channels.

Table 4.5. Incidence of positive sentinel lymph nodes according to tumor thickness

Tumor Thickness (mm)	Patients N	Patients with Positive SLN N	%
< 1.00	123	4	3.2
1.01–2.00	242	25	10.3
2.01–3.00	86	19	22.1
3.01–4.00	69	26	38.5
4.01 +	70	31	44.0

Adapted from Gershenwald et al. Improved sentinel lymph node localization in primary melanoma patients with the use of radiolabeled colloid. Surgery 1998 (In press).

Table 4.6. Prognostic factors influencing SLN pathologic status

Prognostic Factor	Univariate P-Value	Multiple Covariate		
		Odds Ratio	CI	P-Value
Tumor Thickness	< 0.0001	1.23	1.11–1.36	0.0001
Clark Level > III	0.01	1.38	0.83–2.31	0.21
Axial Location	0.17	1.31	0.83–2.09	0.25
Ulceration	< 0.001	2.50	1.53–4.07	0.0002

Abbreviations: CI, 95% confidence interval for odds ratio
Adapted from Gershenwald et al. Improved sentinel lymph node localization in primary melanoma patients with the use of radiolabeled colloid. Surgery 1998 (In press).

patients have had some type of recurrence, and 4% developed failure within the previously mapped regional nodal basin as the sole site or a component of the first site of failure.[21] Such regional nodal basin failures represent another type of false-negative event. Three mechanisms have been proposed to explain nodal failure after a negative sentinel node biopsy:

1) TECHNICAL FAILURE

   The lymphatic-mapping technique did not identify the primary node of drainage and a non-sentinel node was removed rather than the sentinel node. This unidentified sentinel node contained microscopic disease and provided the ultimate source of failure within that basin.

2) PATHOLOGIC FAILURE

   One of the sentinel nodes was removed and contained microscopic disease undetected by conventional histologic techniques. An additional sentinel node or a non-sentinel node with disease remained as the source of subsequent clinical failure.

3) BIOLOGIC OR NATURAL HISTORY-TYPE FAILURE

   At the time of the initial surgery, the correct sentinel node was removed and no metastatic disease was present; however, subclinical intra-lymphatic in-transit disease present since diagnosis provided the ultimate source of subsequent nodal failure.

A more careful histologic examination of the SLNs in patients who developed regional nodal failure as a component of the first site of recurrence after a negative SLN biopsy was also performed. In these studies, the original blocks were subjected to hematoxylin and eosin examination (H&E) of step sections and immunohistochemical staining using antibodies to HMB-45 and S-100. Interestingly, microscopic disease initially undetected by conventional histologic techniques was identified in 8 of the 10 patents who had failed in the nodal basin as a component of the first site of recurrence.[21] Therefore, only two of these nodal failures can be truly classified as false negative events resulting in an actual overall false negative rate of <1%. *The most common mechanism for failure after a negative sentinel node biopsy was actually related to the pathologist's inability to detect microscopic disease using only a conventional approach (H&E staining of bivalved lymph nodes), rather than the surgeon's failure to identify the first node of drainage.* These data therefore provide additional scientific support that the SLN, as defined by these minimally invasive techniques, is the most likely node to contain metastatic disease.

Long-term follow-up of large numbers of patients who underwent SLN biopsy and selective lymphadenectomy has also revealed valuable prognostic information. With a median follow-up of more than three years, prognostic factor analyses of patients from a large database from the M. D. Anderson and Moffitt Cancer Centers demonstrated that the histologic status of the sentinel node was the most powerful predictor of overall survival when compared to previously described primary tumor factors (Table 4.7).[22] Using this information, prognostically

Table 4.7. Univariate and multivariate analyses of prognostic factors influencing disease-specific survival in stage I & II melanoma patients

Prognostic Factors	Univariate P-Value	Multivariate P-Value
Age	.93	.62
Gender	.14	.68
Thickness	< .0001	.02
Clark Level > III	.005	.07
Axial Location	.003	.01
Ulceration	< .003	.33
SLN Status	< .0001	.0001

Adapted from Gershenwald et al. Patterns of failure in melanoma patients after successful lyjmphatic mapping and negative sentinel node biopsy. 49th Annual Meeting of The Society of Surgical Oncology, Atlanta, GA; 1996.

disparate subsets of stage I and II patients can be defined; combined with the availability of effective adjuvant systemic therapy in node positive patients, staging has emerged as a critical motivation for performing a sentinel lymph node biopsy (Fig. 4.2).

ACCURATE NODAL STAGING

The importance of accurate nodal staging is 2-fold:
1) It effectively determines which patients harbor clinically occult nodal metastases and who can therefore benefit from earlier therapeutic node dissections, both from the standpoint of improved local regional control (lymphadenectomy is performed when the tumor burden is low) and potentially improved survival.
2) It identifies those clinical stage I and II patients who are actually occult stage III and who are therefore at higher risk for distant failure. These patients may benefit from earlier administration of systemic adjuvant therapy or may be offered participation in adjuvant therapy trials earlier in the course of their disease. Components of accurate nodal staging include proper and accurate identification of the sentinel nodes as well as careful histologic examination of the nodal specimen. Although the pathologic examinations which define careful histologic examination continue to evolve, it is clear that analysis of one or two special nodes (the sentinel nodes) can be more rigorous compared to the histologic evaluation of all lymph nodes submitted following an ELND.

Historically, the standard approach to the evaluation of SLNs was to bivalve a clinically negative node and stain a section from each half with H&E. This technique probably samples less than 5% of the entire volume of the lymph node and likely represents the most important reason for the underestimation of regional

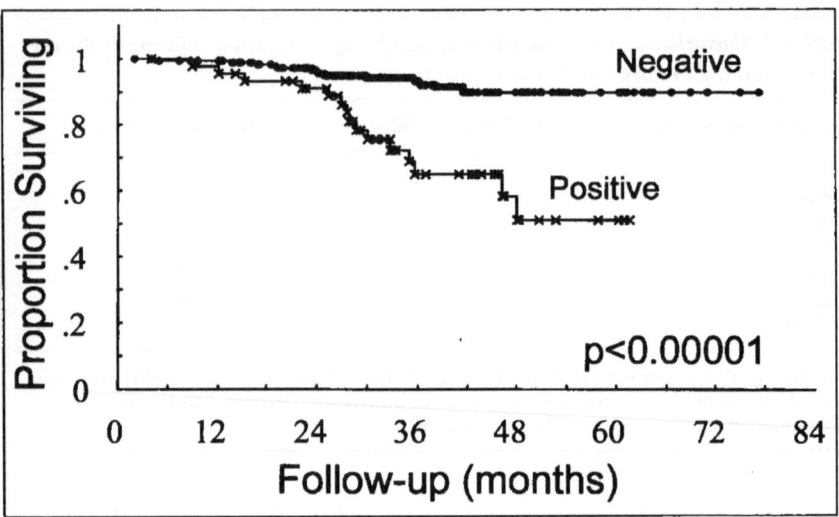

Fig. 4.2. Disease-specific survival. Kaplan-Meier survival for patients undergoing successful lymphatic mapping and SLN biopsy stratified by SLN status. Disease-specific survival was significantly better for patients with a negative SLN biopsy (p < 0.0001).

nodal disease in stage I and II patients. For example, the incidence of nodal failure following surgical excision alone for primary melanomas 1.5-4.00 mm (stage IIa) is approximately 40-50%, while the incidence of microscopic nodal disease in ELND or sentinel node biopsy specimens is approximately one-half that predicted by natural history data alone (Table 4.8).[18,23] While subsequent nodal failure may in part result from microscopic in-transit disease, several lines of evidence support the concept that nodal disease is more often present than is demonstrated by conventional histologic techniques:

1) step sectioning studies (i.e., better sampling) improve the ability to detect microscopic disease,[24-26]

2) 80% of patients who develop nodal basin failure after a negative sentinel node biopsy initially assessed by routine pathology are actually node positive following more careful analysis of the paraffin blocks (see above) and,[22]

3) prospective evaluation of sentinel lymph nodes using the polymerase chain reaction (PCR) to detect the presence of messenger RNA (mRNA) specific for the production of melanoma markers (i.e. tyrosinase) provides results more similar to the natural history data (Table 4.8).[27,28]

The obvious advantage of PCR analysis is that the entire lymph node can be potentially evaluated. The clinical relevance of PCR findings are still under investigation, but preliminary data from the Moffitt Cancer Center suggest that the PCR positive-H&E negative lymph node patients have survival rates intermediate between those patients who are PCR positive-H&E positive and those who are

Table 4.8. Incidence of microscopic nodal disease comparison of histologic technique to natural history

AJCC Stage	Nodal Failure (Natural History)	ELND + H & E	SLN + H & E	SLN + PCR
I (< 1.5 mm)	10–12%	5%	5%	33%
IIa (1.5–4.0 mm)	40–45%	20%	20%	50%
IIb (> 4.0 mm)	60–70%	35%	35–45%	80%

From Reintgen D, Balch C, Kirkwood J, Ross M. Recent advances in the care of the patient with malignant melanoma. Ann Surg 1997; 225:1-14.

Table 4.9. Clinical correlation of nodal status with reverse transcription polymerase chain reaction (RT-PCR)

Nodal Status	Number of Patients	Recurrences	Local	Regional
Histology– RT-PCR +	14	6 (42%)	2	4
Histology– RT-PCR +	27	6 (22%)	3	3
Histology– RT-PCR +	33	2 (6.6%)	1	1

From Reintgen D, Balch C, Kirkwood J, Ross M. Recent advances in the care of the patient with malignant melanoma. Ann Surg 1997; 225:1-14.

both PCR and H&E negative (Table 4.9).[27] Since the volume of material submitted to the pathologist is not only small (1 or 2 nodes), but also the most likely to contain microscopic disease, this approach provides an efficient way to prospectively evaluate the importance of these more sensitive histologic techniques—PCR, step sectioning and immunohistochemical analysis. Such staging information will allow us to define prognostically homogenous subgroups and better stratify patients for adjuvant therapy trials.

CLINICAL TRIALS

LM/SNB provides an opportunity to ask important questions in the context of clinical trials. Such endeavors will greatly impact our understanding of the following issues:

1) the natural history and staging of clinical stage I and II patients,
2) the local-regional control and/or survival benefits for early therapeutic node dissection,
3) the long-term regional nodal basin failure rate following a negative sentinel node biopsy, and

4) the role for systemic adjuvant therapy in patients with subclinical nodal disease.

The Multicenter Selective Lymphadenectomy Trial sponsored by the National Cancer Institute is an ongoing prospective randomized study with a target accrual of 1200 patients. The treatment algorithm for patients with melanomas > 1.0 mm thick is depicted in Figure 4.3. Eligible patients are randomized to LM/SNB versus wide local excision (WLE) alone. Patients with positive sentinel nodes undergo therapeutic lymphadenectomy while sentinel node negative patients are observed. Patients who are initially managed with WLE alone will undergo a therapeutic dissection if clinical regional nodal failure develops. The following questions will be specifically addressed:

1) Is there a survival benefit for patients who undergo early lymphadenectomy?,
2) What, if any, therapeutic value exists for sentinel node biopsy alone?, and
3) What is the false negative rate as determined by failure in nodal basin after a negative SLN biopsy?

Another multi-institutional study—the "Sunbelt Melanoma Trial"—addresses issues related to the relative clinical significance of microscopic nodal disease as determined by different histologic methods:

1) conventional histology,
2) step sectioning and immunohistology, and
3) PCR analysis using four molecular markers (MAGE III, MART I, GP 100 and Tyrosinase). This study will not only elucidate the natural history of these subsets of patients, but will also examine the potential benefit of high dose interferon administered to patients with low nodal burden metastatic disease (Fig. 4.4).

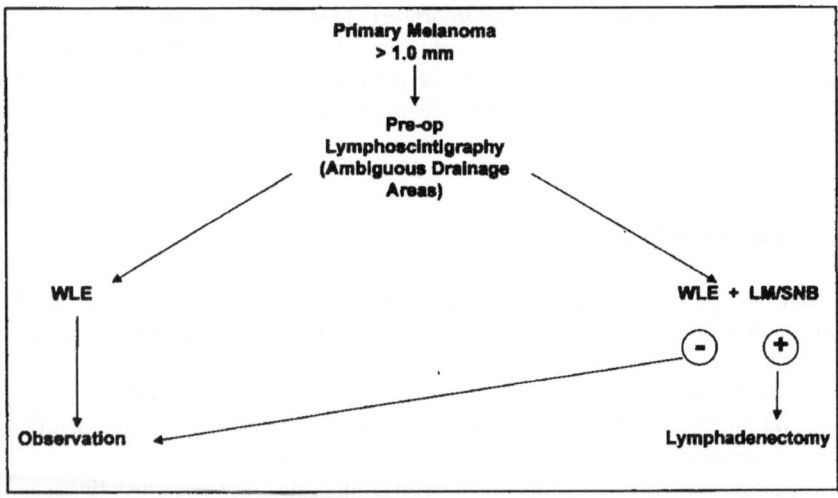

Fig. 4.3. Treatment algorithm for Multi-Center Selective Lymphadenectomy Trial. WLE, wide local excision; LM/SNB, lymphatic mapping/sentinel node biopsy.

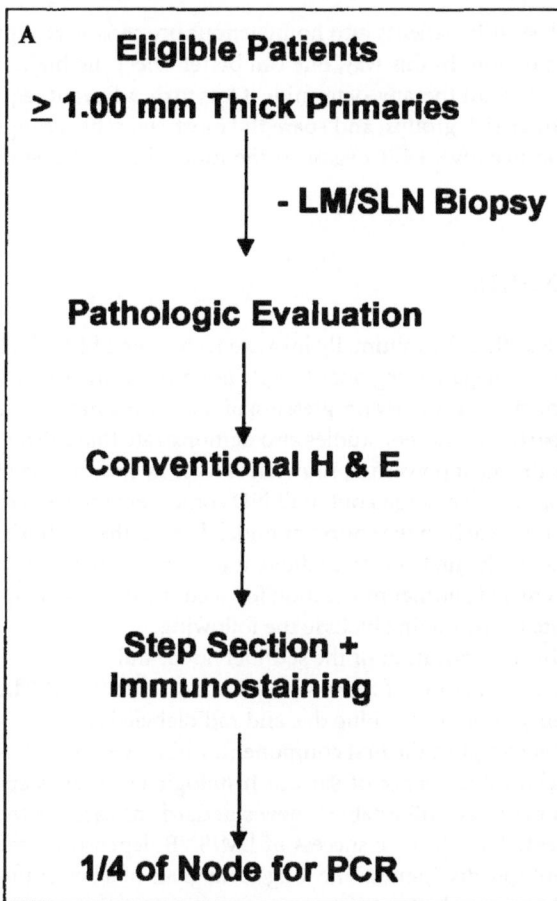

A

Eligible Patients

≥ 1.00 mm Thick Primaries

- LM/SLN Biopsy

Pathologic Evaluation

Conventional H & E

Step Section + Immunostaining

1/4 of Node for PCR

Fig. 4.4. Algorithm for "Sunbelt Melanoma Trial". (A) Techniques for histologic examination of each SLN removed. (B) Treatment according to results of histologic evaluation.

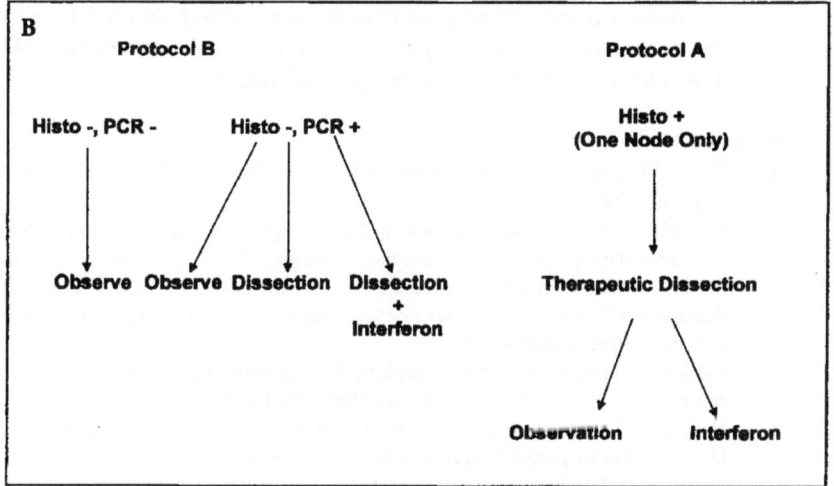

B

Protocol B

Protocol A

Histo -, PCR -   Histo -, PCR +

Histo +
(One Node Only)

Observe Observe Dissection Dissection + Interferon

Therapeutic Dissection

Observation   Interferon

Future clinical trials will stratify patients into homogenous prognostic groups based on sentinel node evaluation. In this way, one can better select the highest risk patients for aggressive adjuvant therapy, determine if less toxic adjuvant regimens may be effective in lower risk groups, and spare the most favorable groups (H&E negative, step section negative, PCR negative) the morbidity and cost of any adjuvant therapy.

CONCLUDING COMMENTS

Numerous studies support that the minimally invasive technique of LM/SNB accurately stages the clinically negative regional lymph node basin in primary melanoma patients and confirms the orderly progression of melanoma metastases within the lymphatic compartment. Recent studies also demonstrate that the histologic status of the SLN is the most powerful prognostic factor in clinical stage I and II patients. While this approach emerged out of ELND controversy as a way to perform therapeutic dissections early in the course of nodal disease, the establishment of high dose interferon as the first effective adjuvant therapy, particularly in node positive patients, has created another motivation for accurate nodal staging. The components of accurate nodal staging include the following:

1) proper and reliable identification of the sentinel node, and
2) careful histologic examination of those sentinel node(s) identified. The combined modality approach of blue dye and radiolabeled colloid injections can best accomplish the first component, while a better understanding of the clinical relevance of various histologic techniques applied to the sentinel node will establish new standards of care for the second component. The ultimate success of LM/SNB depends on the integration of multiple disciplines. The long-term goals of a better understanding of the natural history of melanoma patients and the establishment of more defined guidelines for surgical and adjuvant therapy will be accomplished by using the information obtained from LM/SNB as stratification criteria in the design of clinical trials.

REFERENCES

1. Parker SL, Tong T, Bolden S, Wings, PA. Cancer Statistics, CA: Cancer J Clin 1997; 47:5-27.
2. Ross MI. Surgical management of stage I and II melanoma patients: Approach to the regional lymph node basin. Seminars in Surgical Oncology 1996; 12:394-401.
3. Balch C, Soong S, Bartolucci A et al. Efficacy of an elective regional lymph node dissection of 1 to 4 mm thick melanomas for patients 60 years of age and younger. Ann Surg 1996; 224:255-263.
4. Balch CM. The role of elective lymph node dissection in melanoma: Rationale, results and controversies. J Clin Oncol 1988; 6:163-172.
5. Slingluff CL Jr, Stidham KR, Ricci WM et al. Surgical management of regional lymph nodes in patients with melanoma: Experience with 4682 patients (see comments). Ann Surg 1994; 219:120-130.

6.　Buzaid AC, Tinoco LA, Jendiroba D et al. Prognostic value of size of lymph node metastases in patients with cutaneous melanoma. J Clin Oncol 1987; 39:139-147.

7.　Slingluff CJ Jr, Vollmer R, Seigler H. Stage II malignant melanoma: Presentation of a prognostic model and an assessment of specific active immunotherapy in 1,273 patients. J Surg Oncol 1987; 39:139-147.

8.　Balch CM, Murad TM, Soong SJ et al. Tumor thickness as a guide to surgical management of clinical stage I melanoma patients. Cancer 1979; 43:883-888.

9.　Morton D, Wen D, Wong J et al. Technical details of intraoperative lymphatic mapping for early stage melanoma. Archives Surg, 1992;127:392-399Krag D, Meijer S, Weaver D et al. Minimal-access surgery for staging of malignant melanoma. Arch Surg 1995; 130:654-658.

10.　Kirkwood J, Strawderman M, Ernstoff M, Smith T, Bordern E, Blum R. Interferon-alfa-2b adjuvant therapy of high-risk resected cutaneous melanoma: The Eastern Cooperative Oncology Group Trial EST 1684. J Clin Oncol 1996; 14:7-17.

11.　Reintgen D, Cruse C, Wells K et al. The orderly progression of melanoma nodal metastases. Ann Surg 1994; 220:759-767.

12.　Ross M, Reintgen D, Balch C. Selective lymphadenectomy: Emerging role for lymphatic mapping and sentinel ode biopsy in the management of early stage melanoma. Semin Surg Oncol 1993; 9:219-223.

13.　Thompson J, McCarthy W, Bosch C et al. Sentinel lymph node status as an indicator of the presence of metastatic melanoma in regional lymph nodes. Melanoma Res 1995; 5:255-260.

14.　Berger DH, Feig BW, Podoloff D, Norman J, Cruse CW, Reintgen DS, Ross MI. Lymphoscintigraphy as a predictor of lymphatic drainage from cutaneous melanoma. Ann Surg Oncol 1997; 4:247-251.

14a.　Gershenwald JE, Tseng CH, Thompson W, Mansfield PF, Lee JE, Bouvet M, Lee J, Ross MI. Improved sentinel lymph node localization in primary melanoma patients with the use of radiolabeled colloid. Surgery 1998, in press.

15.　Norman J, Cruse CW, Espinosa C et al. Redefinition of cutaneous lymphatic drainage with the use of lymphoscintigraphy for malignant melanoma. Am J Surg 1991; 162:432-437.

16.　Uren R, Howman-Giles R, Thompson J et al. Lymphoscintigraphy to identify sentinel lymph nodes in patients with melanoma. Melanoma Res 1994; 4:395-399.

17.　Krag D, Meijer S, Weaver D et al. Minimal-access surgery for staging of malignant melanoma. Arch Surg 1995; 130:654-658.

18.　Albertini J, Cruse C, Rapaport D et al. Intraoperative radiolymphoscintigraphy improves sentinel lymph node identification for patients with melanoma. Ann Surg 1996; 223:217-224.

19.　Nathanson SD, Anaya P, Karvelis KC, Eck L, Havstad S. Sentinel lymph node uptake of two different technetium-labeled radiocolloids. Annals of Surgical Oncology 1997; 4(2):104-110.

20.　Kapteijn B, Nieweg O, Liem I et al. Localizing the sentinel node in cutaneous melanoma: gamma probe detection versus blue dye. Ann Surg Oncol 1997; 4(2):156-160.

21.　Gershenwald J, Colome M, Lee J et al. Patterns of recurrence following a negative sentinel lymph node biopsy in 243 patients with stage I or II melanoma. J Clin Oncol 1998; 16:2253-2260.

22.　Gershenwald J, Thompson W, Mansfield P et al. Patterns of failure in melanoma patients after successful lymphatic mapping and negative sentinel node biopsy. 49th Annual Meeting of The Society of Surgical Oncology, Atlanta, GA; 1996.

23.   Schneebaum S, Briele HA, Walker MJ et al. Cutaneous thick melanoma. Prognosis and treatment. Arch Surg 1987; 122:707-711.
24.   Robert MR, Wen DR, Cochran AJ. Pathological evaluation of the regional lymph nodes in malignant melanoma. Semin Diagn Pathol 1993; 10:102-115.
25.   Lane N, Lattes R, Malm J. Clinicopathologic correlations in a series of 117 malignant melanomas of the skin of adults. Cancer 1958; 11:1025-1043.
26.   Das Gupta TK. Results of treatment of 269 patients with primary cutaneous melanoma: a five-year prospective study. Ann Surg 1977; 186:201-209.
27.   Reintgen D, Balch C, Kirkwood J, Ross M. Recent advances in the care of the patient with malignant melanoma. Ann Surg 1997; 225:1-14.
28.   Wang X, Heller R, VanVoorhis N et al. Detection of submicroscopic lymph node metastases with polymerase chain reaction in patients with malignant melanoma. Ann Surg 1994; 220(6):768-774.

Sentinel Lymph Node Biopsy for Melanoma: Surgical Technique

Fadi F. Haddad, Douglas Reintgen

5

INTRODUCTION

The care of patients with melanoma has changed in the last 5 years with the development of new lymphatic mapping techniques to reduce the cost and morbidity of nodal staging, the emergence of more sensitive assays for occult melanoma metastases, and the identification of interferon alfa-2b as an effective adjuvant therapy for the treatment of patients with melanoma at high risk for recurrence. The accurate staging of melanoma patients has become more important in light of the recent report of a multi center, prospective, randomized trial that shows a survival benefit for patients with T4 (tumor thickness > 4.0 mm) or Stage 3 (nodal metastases) melanoma who are treated with adjuvant interferon alfa-2b (Intron A®, Schering Corporation, Kenilworth, New Jersey).[1] The lymphatic mapping technology is the least morbid and costly method to obtain nodal status of the patient with melanoma. Surgical technique is important, but it must be emphasized that the surgeon needs excellent nuclear medicine and pathology support to perform this technique.

THE ROLE OF NUCLEAR MEDICINE—SURGICAL PERSPECTIVE

Lymphoscintigraphy has been shown to be indispensable in predicting lymphatic basins at risk for the development of metastatic disease in patients with cutaneous malignant melanoma.[2] Preoperative lymphoscintigraphy serves as a road map for the surgeon planning the surgical procedure in the following ways:

Radioguided Surgery, edited by Eric D. Whitman and Douglas Reintgen.
© 1999 Landes Bioscience

Fig. 5.1. 50 year old white male with an "intermediate thickness" melanoma on the right shoulder. Preoperative lymphoscintigraphy with Technetium Sulfur Colloid shows bidirectional drainage to sentinel nodes in both the left neck (open arrow) and left axilla (closed arrow). Both basins are at risk for metastatic disease, while the clinical prediction would be for drainage to just the left neck. Intraoperative lymphatic mapping harvested the SLN in both locations with the neck being negative and the axilla having micro metastatic disease.

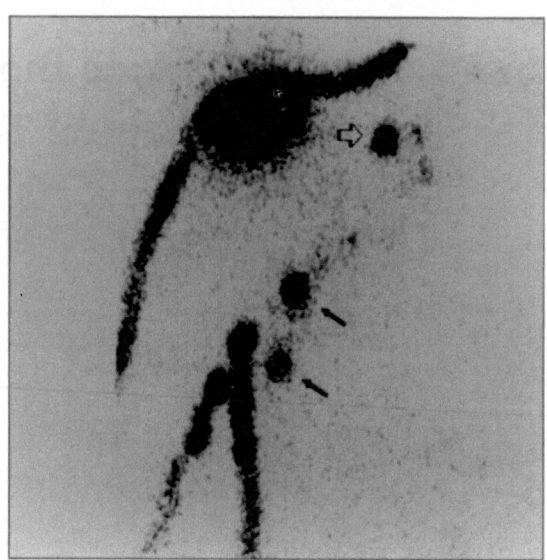

1. To identify all nodal basins at risk for metastatic disease (Fig. 5.1).

2. To identify any in-transit nodes that can be tattooed by the nuclear medicine colleague for later harvesting. In-transit metastases occur in 5% of the melanoma population and may, by definition, be considered the SLN (Fig. 5.2).

3. To identify the location of the SLN in relation to the rest of the nodes in the basin.[3] The location of the SLN may be variable in a basin and ideally the surgeon needs a mark of the position of the SLN in reference to other nodes in the basin, in order to perform the harvest under local anesthesia with a minimal incision. Preoperative lymphoscintigrams can do this quite well. Twenty-nine patients with clinically negative nodes and melanomas thicker than 0.76 mm had preoperative lymphoscintigrams in two planes to mark the location of the SLN prior to operation. Thirty-three percent of the time the clinician could not predict the approximate location of the SLN within 5 cm, but the lymphoscintigraphy was accurate in the identification of the location of the SLN within 1.0 cm 100% of the time.[3] The technique was most accurate in the groin and the head and neck where the lymph nodes are more superficial. The axilla is the most difficult area to map and the best the preoperative lymphoscintigram can do in this basin is to tell the surgeon whether the node is located anterior, posterior, superior or inferior in the basin.

4. To estimate the number of SLN in the regional basin that will need to be harvested.

Preoperative lymphoscintigraphy is a simple, accurate test that is the most efficient, cost-effective method of identifying the lymph nodes at highest risk for metastatic disease.

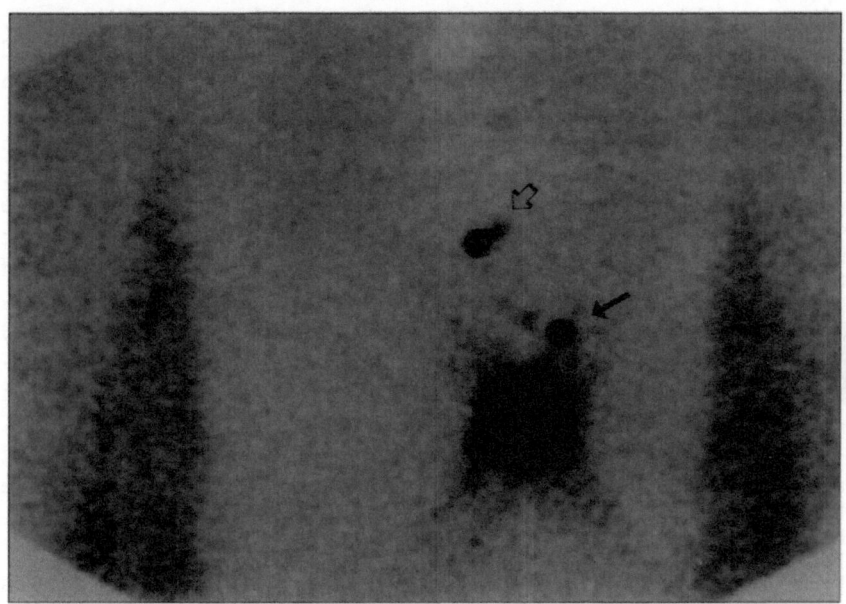

Fig. 5.2. 35 year old white male with a 2.0 mm malignant melanoma excised from the left upper abdomen. Clinically the predicted cutaneous lymphatic flow would be to the left axilla. With preoperative lymphoscintigraphy, the left axilla is found not to be at risk for metastatic disease, but lymphatic flow was noted to an in-transit node in the left upper abdomen within 5 cm of the primary site (closed arrow) and to a lower internal mammary node (open arrow). Both nodes were mapped intraoperatively and biopsied. Pathology was negative.

DESCRIPTION OF THE TECHNIQUE OF INTRAOPERATIVE LYMPHATIC MAPPING

There is great variation from center to center on the technique of intraoperative mapping. This section will discuss the nuances of our technique, detailing the steps we believe important for successful mapping. However, similar success rates are possible with slight variations in technique, as shown by multiple centers.

Patients are scheduled in the nuclear medicine suite early the day of the surgery and undergo preoperative lymphoscintigraphy with the injection of technetium around the primary site (see chapter 8 for details). We do not obtain lymphoscintigraphy before the day of surgery; this is redundant and costly. Dynamic scans are performed 5-10 minutes after the injection of the radiocolloid and the location of the SLN is marked in the basin with an intradermal tattoo. Delayed images (2 hours) for trunk and head and neck melanoma may be necessary to ensure there is not delayed drainage to more than one regional basin.

The patient is then taken to the OR 2-6 hours after radiocolloid injection and 1 cc of 1% lymphazurin (isosulfan blue) is injected intradermally (per number of directions of drainage, i.e., 1 cc of dye per lymphatic basin identified by scan) around the primary site. After prepping and draping the primary site and regional basin and allowing 10 minutes for the vital blue dye to travel to the SLN, attention is directed initially to the regional basin. With a hand-held gamma probe (e.g., Navigator™, United States Surgical Corporation, Norwalk, CT), the "hot" spot in the regional basin is identified and the hot spot/background ratio is noted. If "shine through" from the primary site is a problem, the wide local excision (WLE) of the primary may be performed first. An incision is made over the hot spot of the sentinel node and small flaps are created in all directions to allow identification of the blue stained afferent lymphatics. Surgical dissection is aided by both visualization of the stained afferent lymphatic down to the blue-stained node (Fig. 5.3) and by the use of the hand held gamma probe. At times, the surgeon can be confused as to what is proximal or distal on the afferent lymphatic and the probe can be used to identify the direction of the dissection. The SLN is identified and removed with sharp or electrocautery dissection. After removal of the entire SLN, afferent and efferent lymphatics from the SLN are controlled with hemoclips, since the electrocautery does not seal lymphatics. This technique decreases the chance of postoperative wound seroma.

Lymph nodes are confirmed in vivo and ex vivo to be SLN with gamma probe, by computing localization ratios. In vivo radioactivity of the suspected SLN is compared to non-SLN tissue. The ratio between these two numbers should be at least 3x. After removal of a SLN, the ex vivo measured cps of the SLN compared to excised non-SLN tissue should be at least 10x. For blue-stained lymph nodes, localization ratios confirm them as SLN by each mapping technique. Non-blue stained lymph nodes are identified as SLN by their localization ratios.

Studies[4,5] have shown that the localization ratios double if the harvest occurs 2-6 hours after the injection of the radiocolloid compared to performing the mapping immediately after the injection of the radiocolloid. This data supports the timing intervals we use, to maximize localization ratios and hence facilitate SLN identification. After removal of the SLN(s), measured radioactivity in the lymphatic basin should decrease, without persistent focal "hot spots." Dissection and identification of more SLN continues until this global decrease is documented.

The use of the radiocolloid for intraoperative mapping allows for excision of the SLN in unusual locations, such as the in-transit nodes. Figure 5.2 illustrates the preoperative lymphoscintigraphy of a patient with an intermediate thickness melanoma of the left abdominal wall. Experience would predict drainage to the left axilla, but preoperative lymphoscintigraphy shows drainage to an in-transit node located within 5 cm of the primary site and further drainage noted to a lower left internal mammary lymph node.

The radiocolloid and vital blue dye mapping techniques are complimentary and we believe they should be used simultaneously to increase the success rate of localization of the SLN. The different mapping techniques are important depending on the location of the primary in relation to the regional basin. If the primary site is close, overlying or in a direct line to the basin so that a hand held gamma probe will detect "shine through" from the residual radioactivity at the primary

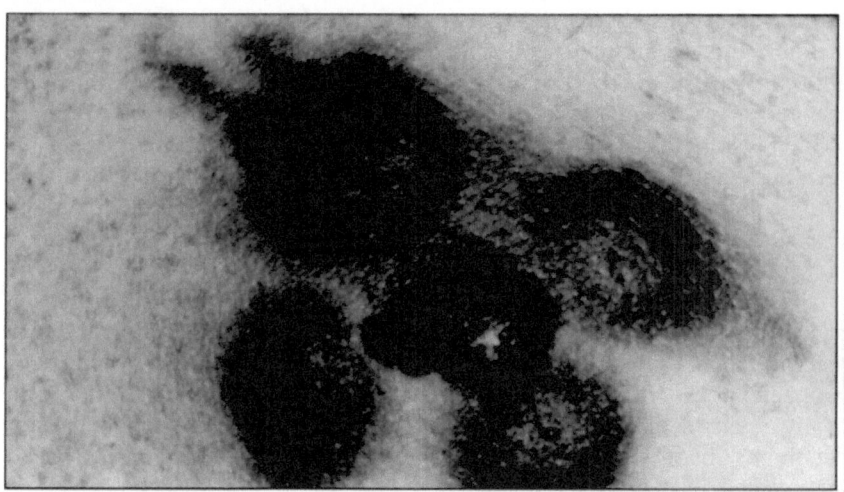

Fig. 5.3. (See Color Insert for color representation). Colored radiocolloid injection into the dermis of the skin in 4 quadrants around a nodular melanoma. Within minutes of the injection, tracer is seen being taken up by the cutaneous lymphatics. It is rare to have the primary lesion intact at the time of the mapping, and most of the data generated for success rates for melanoma mapping is after an excisional biopsy.

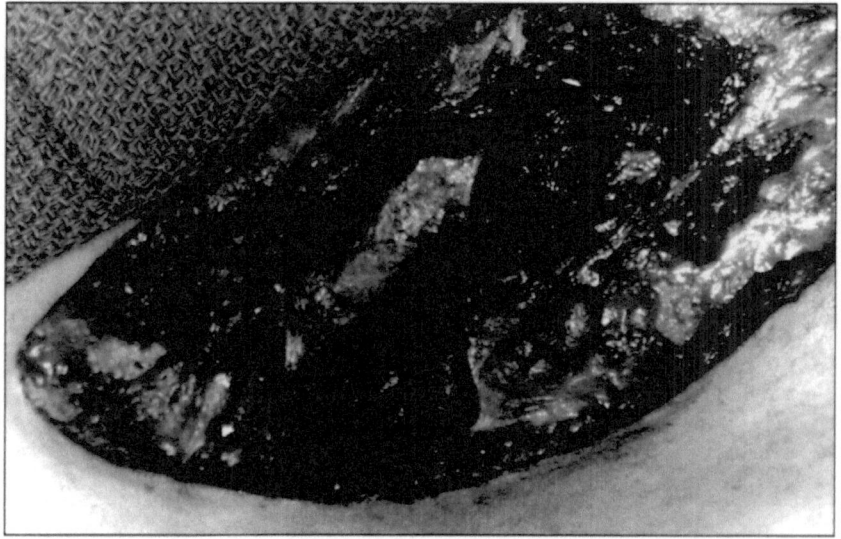

Fig. 5.4. (See Color Insert for color representation). A blue-stained afferent lymphatic is shown entering a blue-stained node. This node will be hot with the gamma probe, will be the SLN and will be the first site of metastatic disease.

site, use of the vital blue dye may be the only technique that allows for successful mapping. Even performing the WLE first, enough radioactivity may still be present at the primary site. In contrast, in patients with a fatty axilla or in head and neck mapping, it may be impossible to follow a wisp of blue-stained afferent lymphatic to the SLN. Particularly in the head and neck area, because of the presence of surrounding vital structures, large flaps are to be avoided and the ability of the gamma probe to locate the "hot" spot through the skin is a tremendous advantage, since it directs both the location of the (smaller) incision and also the dissection vector, once the incision is made.

SLNs are defined as those nodes either in an in-transit location or in the major regional basin that initially and/or primarily receive lymphatic flow from the primary melanoma site. They can be identified by following a blue-stained afferent lymphatic to a blue stained node (blue node), as originally described by Morton[6] or with a gamma probe determined localization ratio. Using these techniques and definitions, the success rate of SLN mapping and biopsy should approach 100% for melanoma patients.[4]

ROLE OF PATHOLOGY

After harvesting, the SLN is then submitted for a detailed histologic examination that may include serial sectioning, immunohistochemical staining with S-100 and/or HMB-45 monoclonal antibodies, and perhaps newer assays using molecular biology techniques for occult metastases.[7] See the chapter 10 for further details.

INCLUSION AND EXCLUSION CRITERIA

It is evident that lymphatic mapping technology has changed the standard of surgical care for melanoma. A crucial issue is which patients should or should not be offered the procedure. We use the following indication for SLN mapping:

1. Cutaneous malignant melanoma with thickness 0.76 mm or greater. (Stage Ib or above)
2. Ulcerated melanomas
3. Thinner lesions may be considered if the melanoma had three or more of the following characteristics: a) located on the trunk, head or neck, b) male patients, c) the lesion is Clark level III or greater, d) regression on histopathologic examination. These variables may indicate that the melanoma is a more aggressive lesion or at one time was a thicker lesion.

Our exclusion criteria are:

1. Clinically positive regional nodal basins or evidence of in-transit, distant skin or systemic metastases.
2. Prior wide local excision (relative), particularly on the head and neck and trunk areas. The data on this are unclear.
3. Previous Z-plasty or rotational flap. This primary site reconstruction, particularly in the head and neck, is too likely to disrupt the cutaneous

lymphatics, and is not reflective of cutaneous lymphatic flow when the primary was intact.

4. Preoperative lymphoscintigraphy showing drainage to more than three basins (relative). The question would arise whether this is too extensive a staging surgical procedure to perform on an elective basis.

5. Other medical problems (relative). However, it is unlikely that a patient's other medical problems would be severe enough to prohibit the performance of a low morbidity procedure such as SLN biopsy.

6. Patients with an allergy to isosulfan blue or sulfur colloid (extremely rare).

Patients with melanoma of tumor thickness between 0.76-1.0 mm (5.3% chance of harboring metastatic disease) are given the choice of having the SLN harvested vs observation. Invariably, patients elect to have the SLN biopsy because the morbidity of the procedure is low and the treatment decisions if one has metastatic disease in the regional basin is radically different than if the SLN is negative. Likewise, patients with thick melanomas (greater than 4.0 mm in tumor thickness) are recommended to undergo lymphatic mapping for staging since T4N0 patients have a better survival than T4N1 patients. In addition, even though interferon alfa (Intron A®) is approved for patients with T4N0 disease, that subgroup was small and showed no benefit in the randomized trial. Many medical oncologists are hesitant to treat this subgroup with the toxic Intron A®. Ongoing national trials are addressing this question. In the meantime, it is reasonable to perform lymphatic mapping and SLN harvest on patients with melanomas greater than 4.0 mm and if nodal disease is found, consider them for adjuvant therapy. Overall, when given a choice, almost all patients will choose to "know" rather than "watch and wait," especially given the option of the low morbidity SLN procedure.

Patients with a positive SLN should undergo a completion lymph node dissection (CLND). This is considered the standard of care for all patients with metastatic disease in any SLN. However, a high percentage of patients will have no further disease within the nodal basin (i.e., in the non-SLN). This phenomenon, termed "exclusivity" elsewhere in this handbook, may occur in more than 90% of patients, with at least one positive SLN. This finding would seem to validate the principle that the SLN is the primary draining lymph node of the tumor site, and that tumor cells (and mapping agents) may be "trapped" within that SLN for a period of time before migrating to other lymph nodes.

The important question of whether one can do lymphatic mapping after a wide local excision was examined retrospectively at our institution. One hundred and seventy patients underwent lymphatic mapping and SLN biopsy after the WLE of the primary site, and 399 patients with melanoma underwent mapping prior to the WLE. The patients that had the procedure after the WLE had an increased incidence of "skip" metastases, had an increased number of SLN removed/basin (1.7 vs 1.9 SLN/basin) and had a significantly increased number of multiple (26% vs 15%, p < 0.05) basins. These data suggest that WLE may disrupt and/or alter cutaneous lymphatic flow, rendering subsequent SLN mapping less effective. This is a controversial area; pathologically positive SLN can be located after WLE, and as SLN technology is adopted around the country, many patients will be referred to SLN-experienced surgeons after WLE.

Radioguided SLN mapping minimizes surgical dissection by directing the operator to the site of the lymph node in vivo. This permits the use of smaller incisions and facilitates biopsy of deep nodes, such as in the axilla. Also, SLN in previously poorly accessible areas can be identified and biopsied, including scapular, internal mammary, and intra-parotid nodes.

Complications of the procedure are few. Drains are not placed, and in follow-up there is a 10% incidence of seroma formation that can be handled easily with aspiration. The side effects of the vital blue dye is staining of the tissue (up to a period of 12 months) if all the blue dye is not removed with the WLE. This does not seem to be a problem in the head and neck area probably due to the rich lymphatic and vascular supply in this area. The urine turns blue/green for 24-48 hours and patients should be informed of this likelihood. Clinicians should be aware that the oxygen saturation reading on the pulse oximeter may decrease secondary to the blue dye. Rarely, an allergy to the blue dye (with hives and a blue discoloration of the skin) may be encountered.

UPDATED RESULTS OF LYMPHATIC MAPPING FOR MELANOMA FROM OUR INSTITUTION

A total of 693 patients with melanoma have undergone lymphatic mapping and SLN biopsy at our institution, Moffitt Cancer Center. The SLN was successfully identified and harvested in 688 patients, for a 99% success rate. One hundred patients (14.5%) had evidence of nodal metastases. The rates of SLN involvement for primary tumors < 0.76 mm, 0.76 mm-1.0 mm, 1.0-1.5 mm, 1.5-4.0 mm and greater than 4.0 mm in thickness was 0%, 5.3%, 8%, 19% and 29%, respectively. Eighty-one patients underwent a completion lymph node dissection (CLND) after positive SLN biopsy and only six patients (7.4%) with positive SLN demonstrated metastatic disease beyond the SLN. The tumor thickness of these 6 patients ranged from 2.8-6.0 mm. No patients with a tumor thickness less than 2.8 mm was found to have evidence of metastatic disease beyond the SLN upon CLND. It is hypothesized that a melanoma has to reach a certain thickness before it sheds off enough cells to involve more than the SLN with disease. This hypothesis was recently supported by data from MD Anderson that showed no non-SLN involved with metastatic disease after a positive SLN biopsy for melanomas less than 2.5 mm in thickness.[11]

CONCLUSION

Radioguided surgery may be the next revolution in general surgery, following in the footsteps of the laparoscopic procedures a decade ago. With potential applications to colon and breast cancers, other skin tumors like the Merkel cell carcinoma or poorly differentiated squamous cell carcinomas, parathyroid localizations, bone tumor localizations and vulvar or vaginal lesions, there are 350,000 new cases of cancer diagnosed each year in the United States to which one can apply the techniques. For melanoma, extensive data from multiple institutions confirm that the SLN are the lymph nodes at highest risk for metastatic disease; that these nodes may successfully be mapped preoperatively with lymphoscintigraphy; and, utilizing state-of-the-art technology, the SLN may be located intraoperatively.

REFERENCES

1. Kirkwood JM, Strawderman MH, Ernstoff MS, Smith TJ, Borden EC, Blum R. Interferon Alfa-2b adjuvant therapy of high-risk resected cutaneous melanoma: The Eastern Cooperative Oncology Group Trial EST 1684. J Clin Oncol 1996; 14:7-17.

2. Norman J, Cruse CW, Espinosa C et al. Redefinition of cutaneous lymphatic drainage with the use of lymphoscintigraphy for malignant melanoma. Am J Surg 1991; 162:432-437.

3. Godellas CV, Berman C, Lyman G et al. The identification and mapping of melanoma regional nodal metastases: minimally invasive surgery for the diagnosis of nodal metastases. Am Surg 1995; 61:97-101.

4. Albertini JJ, Cruse CW, Rapaport D et al. Intraoperative radiolymphoscintigraphy improves sentinel lymph node identification for patients with melanoma. Ann Surg 1996; 223:217-224.

5. Nathanson SD, Anaya P, Eck L. Sentinel lymph node uptake of two different radionuclides. The Society of Surgical Oncology, 49th Cancer Symposium, Atlanta, March, 1996 (Abstract).

6. Morton DL, Wen DR, Cochran AJ. Management of early-stage melanoma by intraoperative lymphatic mapping and selective lymphadenectomy or "watch and wait." Surgical Oncology Clinics of North America 1992; 1:247-259.

7. Wang X, Heller R, VanVoorhis N et al. Detection of submicroscopic metastases with polymerase chain reaction in patients with malignant melanoma. Ann Surg 1994; 220:768-774.

8. Reintgen DS, Cruse CW, Berman C, Ross M, Rapaport D, Glass F, Fenske N, Messina J. An orderly progression of melanoma nodal metastases. Ann Surg 1994; 220:759-767.

9. Ross M, Reintgen DS, Balch C. Selective lymphadenectomy: Emerging role of lymphatic mapping and sentinel node biopsy in the management of early stage melanoma. Seminars in Surgical Oncology 1993; 9:219-223.

10. Gershenwald J, Thompson W, Mansfield P. Lee J, Colome M, Balch C, Reintgen D, Ross M. Patterns of failure in melanoma patients after successful lymphatic mapping and negative sentinel node biopsy. Society of Surgical Oncology, 1996 (Abstract).

11. Haddad FF, Stall A, Messina J, Brobeil A, Ramnath E, Glass F, Cruse CW, Berman C, Reintgen DS. The orderly progression of melanoma nodal metastases is dependent on tumor thickness of the primary lesion. Ann Surg Oncol, Submitted.

Technique for Lymphatic Mapping in Breast Carcinoma

Charles E. Cox

INTRODUCTION: ECONOMIC IMPERATIVES

The surgical treatment of breast cancer is rapidly evolving. Some factors influencing these changes include altered disease demographics, advances in technology, governmental and reimbursement controls, and increasing public expectations. The epidemiology of breast cancer as "baby boomers" turn 50 years of age marks a dramatic rise in the prevalence of breast cancer. Although the relative risk of developing breast cancer has remained constant, the incidence of breast cancer is projected to increase from 185,000 new cases annually to 420,000 new cases annually over the next 20 years. Advances in surgical technology and the development of minimally invasive surgical techniques have heralded a new era in surgery.[3] The new technology of sentinel lymph node (SLN) mapping for breast carcinoma, as with any other new surgical technique in this era, must meet the burden of not only improved efficiency and reduced risk but also diminished cost and resource utilization.[4] The added burden of economic (reimbursement) pressure and increased public expectations for better cancer therapies place additional scrutiny on surgeons utilizing these new techniques.

I believe that safe implementation of SLN biopsy for breast cancer requires two things: adequate training programs and initial performance only under an approved IRB protocol. A recent publication introduced several guidelines regarding these issues.[5] Training, credentialing, IRB protocols and outcome measures are discussed in other chapters of this book. Our program has developed our own flow-tech, coded and secure forms for recording outcomes. These forms can be faxed to a secure internet repository <*http://mapping.rad.usf.edu*> allowing surgeons to enter, review and monitor their own outcomes.[5] We are excited about

Radioguided Surgery, edited by Eric D. Whitman and Douglas Reintgen.
© 1999 Landes Bioscience

this new program, because it will provide confidential access to multi-institutional clinical research data to users around the world.

STATUS OF AXILLARY NODE DISSECTION: HISTORICAL OVERVIEW

The current standard of care for the management of invasive breast cancer is the complete removal of the tumor and documentation of negative margins by either mastectomy or lumpectomy followed by complete axillary lymph node dissection.[6-8] Lymphatic mapping of the breast is clearly changing this long held paradigm.

The role of axillary dissection may be currently the most controversial topic in the treatment of breast cancer. Nearly 100 years ago, Halsted demonstrated the curative potential of radical mastectomy. Fifty years later, Patey proved that modified radical mastectomy could yield similar survival with limited morbidity. The controversy now rages over the current role of axillary dissection in the management of operable breast cancer. Since the time of Halsted, the status of the regional nodal basin remains *the single most important independent variable of predicting prognosis.*[9] Advocates of axillary dissection contend that there is a benefit for breast cancer patients since axillary dissection renders regional control of axillary disease. They propose that surgical removal of microscopic nodal metastases is curative without adjuvant chemotherapy in certain patient populations. Nevertheless, recent critics of axillary dissection maintain that overall survival depends on the development of distant metastases and is not influenced by axillary dissection in most patients.[10-12] They contend that patients with microscopic axillary metastases might be cured with adjuvant chemotherapy with or without nodal irradiation without the need for axillary dissection. Many have even advocated the abandonment of axillary dissection in early breast cancer. Sentinel node biopsy for breast cancer may eliminate this controversy. This low morbidity procedure appears to obviate the need for routine axillary node dissection because it correctly identifies those patients with nodal metastases who require completion node dissection while also defining that subset of patients without any nodal disease who are unlikely to benefit from further node removal.

HISTORICAL PERSPECTIVE OF THE DEVELOPMENT OF LYMPHATIC MAPPING FOR BREAST CANCER

Lymphatic mapping was initially developed for the treatment of malignant melanoma and has subsequently been adopted for breast cancer lymph node evaluation.[14-18] There are obvious differences between melanoma and breast cancer. Melanoma generally is faster growing and declares itself quickly. Therefore, the magnitude of the delay for clinical detection of recurrent nodal metastases following surgical and histologic examination of sentinel nodes for metastatic breast cancer could take between 3-5 years versus the 8-18 months required for the same clinical detection of melanoma metastases. Furthermore, these delays could be further prolonged by the broad application of adjuvant therapies for breast cancer in node negative patients, especially if detection of lymph node metastases is minimized during the initial surgical evaluation. The possibility also exists, as with melanoma, that some patients may be spared the use of adjuvant chemotherapy if no nodal metastases can be found with these highly sensitive techniques. (See elsewhere in this handbook for details on pathologic examination.)

The purpose of this chapter is to describe the current technique for lymphatic mapping of breast carcinoma. Covered in this section will be a current description of the methods to perform lymphatic mapping using both blue dye and radiocolloid. Included will be sections on: the characteristics of both blue dye and radiocolloid and injection techniques for both, operative techniques, gamma mapping techniques and finally a section on intraoperative measurement recording and documentation of results.

TECHNICAL ASPECTS OF LYMPHATIC MAPPING FOR BREAST CARCINOMA

RADIOCOLLOID INJECTION

In the United States, the radiocolloid employed for SLN mapping is technetium sulfur colloid, although other compounds are used elsewhere in the world.[19] For example, antimony sulfide, no longer available in the U.S., has successfully been used for SLN mapping in Australia and other countries.

In some of the initial pilot studies of mapping at our institution, subdermal injections were used due to the richness of the subdermal plexus of lymphatics. However, though mapping was excellent it was unclear if this is a true representation of the lymphatic flow of the tumor bed. We abandoned this method due to this uncertainty. Guiliano's work and our current series of 700 mapping cases have less than a 1% skip rate which would argue strongly that intraparenchymal injection of the mapping agents is the optimal route.[20] This low skip rate is consistent with the clinical imperative of SLN mapping for breast cancer: that the SLN be pathologically representative of the nodal metastatic status of the patient. It has been our observation that deep parenchymal injections of blue dye have appeared in the skin with time. Does the lymph flow to the skin then to the nodes in some cases? Are deep lymphatics draining to different nodes than skin lymphatics? Further studies perhaps using PET scanning may more carefully elucidate the actual lymphatic flow of the breast and definitively resolve the question of where mapping agents should be injected.

All studies performed within the U.S. employ Tc^{99m} labeled sulfur colloid either in a filtered or unfiltered state injected in the amount of 0.450 mCi-1.0 mCi of specific activity bound to the sulfur colloid respectively. Krag and his group have routinely used unfiltered Tc^{99m} labeled sulfur colloid and injected 1.0 millicuries of radioactivity. We utilize 450 microcuries of 0.22 micron filtered sulfur colloid, (in combination with intraoperative Lymphazurin Blue dye). The radio-nuclide is injected in a 6 cc volume at six different sites intraparenchymally surrounding the tumor or the tumor bed following excisional biopsy.

BLUE DYE TECHNIQUE

There are two agents referred to in the mapping literature trademarked as Lymphazurin and Patent blue dye. Essentially, these are biochemically the same agents. Lymphazurin 1% (isosulfan blue) is a sterile aqueous solution for subcutaneous administration. Phosphate buffer and sterile, pyrogen-free water is added in sufficient quantity to yield a final pH of 6.8 -7.4. Each cc of solution contains 10 mg isosulfan blue, 6.6 mg sodium monohydrogen phosphate and 2.7 mg potassium bihydrogen phosphate. The solution contains no preservatives. It is a contrast agent for the identification of lymphatic vessels without known pharmacologic action. Following subcutaneous administration isosulfan blue is selectively picked up by the lymphatic vessels, giving them a bright blue-color discernible from surrounding tissue. There is some evidence that 50% of the isosulfan blue, from aqueous solution, is weakly bound to serum protein (albumin). Since interstitial protein is presumed to be carried almost exclusively by lymphatics and in view of evidence of binding of dye to proteins, visualization may be due to protein binding phenomenon. Up to 10% of the subcutaneously administered dose of Lymphazurin 1% is excreted unchanged in the urine in 24 hours in man. Presumably, 90% is excreted through the biliary route.

Lymphazurin 1% (isosulfan blue) has demonstrated a 1.5% incidence of adverse reactions. All the reactions were of an allergic type. Localized swelling at the site of administration and mild hives of hands, abdomen and neck have been reported within several minutes following administration of the drug. Reports of mild to severe reactions have appeared in the literature for compounds similar to isosulfan blue. There have been no reported deaths from the administration of isosulfan blue dye. However, a death has been reported following intravenous administration of a similar compound employed to estimate the depth of a severe burn. Severe reactions may be manifested by edema of the face and glottis, respiratory distress or shock. Such reactions may prove fatal unless promptly controlled by such emergency measures as maintenance of a clear airway and immediate use of oxygen and resuscitative drugs.

To date in our current series of over 700 patients, approximately 1% have demonstrated allergic reactions to isosulfan blue. These have been manifested by an initial wheal reaction at the injection site followed by the development of blue hives scattered about the neck, groin, and other intertriginous areas. These have generally responded to IV Benadryl, and have disappeared rapidly. It is advisable to observe the patient at least 30-60 minutes following the administration of isosulfan blue since severe or delayed reactions may occur.

The admixture of Lymphazurin with local anesthetics, i.e., lidocaine, in the same syringe prior to administration results in an immediate precipitation of 4-9% of the drug complex. Similar precipitation of sulfur colloid occurs with the mixture of these two compounds. This technique is not recommended. Is in the best interest of the patient to give other medications such as local anesthetics or sulfur colloid via separate syringe and at separate time intervals.

Lymphazurin has no known carcinogenic, mutagenic, or teratogenic effects but should be given to pregnant women only if clearly needed. Safety and effectiveness of Lymphazurin for children has not been established. It is unknown whether the drug is excreted in human milk and should be administered with caution to nursing mothers.

INJECTION TECHNIQUE FOR LYMPHATIC MAPPING WITH BLUE DYE

There are occasional reports of lymphatic mapping for breast carcinoma with intradermal injection of blue dye. The same high percentage uptake in lymph nodes occurs due to the rich subdermal lymphatic channels; however, the same arguments apply to this technique as are described above with the intradermal injection of sulfur colloid. The authors contributing to this handbook recommend intraparenchymal, not the intradermal, injection of Lymphazurin for the lymphatic mapping of breast carcinoma.

In our preferred method, an intraparenchymal injection of 5 cc of blue dye is performed in multiple sites around the tumor or excisional biopsy site. For lesions in the upper outer quadrant of the breast which assuredly drain to the axilla, the dye may be injected through a 27 gauge needle and fanned through a single injection site into the upper outer aspect of the breast towards the axilla. Following this injection, manual compression of the breast and gentle massage for a sustained time of five minutes by the clock is performed to insure proper migration of the blue dye into the lymphatic channels. This is performed just prior to skin preparation and draping of the patient for operative intervention. Immediately following this preparation of the patient the sentinel lymph node can be found by making an accurate incision approximately 1 cm inferior to the hair line of the axilla. Dissection may proceed quickly to the clavi-pectoral fascia after which care must be taken to avoid damage to any lymphatic channels seen beyond that point. Disruption of the lymphatic channels at this level will preclude the ability to find a SLN. Channels may appear superficial to this area and may lead to the SLN; however, other channels at a deeper level are more likely to carry the majority of the blue dye to the SLN.

OPERATIVE TECHNIQUE FOR LYMPHATIC MAPPING
OF BREAST CARCINOMA

In the ideal circumstance the patient would have been injected at least 2 hours previously with 450 microcuries of filtered Tc^{99m} labeled sulfur colloid. Lymphazurin 1% (isosulfan) blue dye, 5 cc would have been injected, compressed for five minutes and the skin of the breast, chest wall and axilla would have been prepared for operation with appropriate scrubs and disinfectant solutions (Hibiclens and Hibistat respectively preferred by the author). The patient is appropriately draped to expose the breast, chest wall, and axilla. The gamma probe to be utilized is sterilely sheathed and available for axillary node mapping.

Before proceeding to the actual mapping, it is important to mention several general considerations. Since most surgeons are just learning the new technique, they will experience an initial learning curve. Since new operative techniques may require additional time, it is important that the patients be scheduled to allow enough time for the operation to proceed unencumbered by time constraints. For example, most surgeons can easily perform axillary dissection quicker than they will be able to do their initial sentinel lymph node biopsies. The procedure should be done in a calm environment and it is encouraged as one begins the mapping and selective lymphadenectomy that the operator and assistant sit down to operate.

Visualization of small lymphatic channels requires good exposure and retraction in a bloodless field. Therefore, is encouraged that one uses the electrocautery for dissection with a self-retaining retractor in the incision and if using local anesthetic, epinephrine should be included. Careful dissection is mandatory and the author has found the use of small fine tipped clamps (mosquito, classic delicate, 6", Codman 30-4470) for dissection to be extremely useful. One should proceed with caution. Lymphatic channels should be clipped. However, it is imperative that one not cut nor clip the blue channel containing the dye until a node or nodes have been isolated. Do not cut the blue channel, as this will result in loss of drainage of the blue dye into the SLN, making it very difficult to localize without the gamma probe. Furthermore, the blue dye will stain the surrounding tissues further complicating the ability to detect the blue lymph node. If inadvertently the blue dye channel is cut then the use of the gamma detection probe is critical in location of the SLN.

Mapping the axillary SLN requires a properly calibrated machine in which the sensitivity is properly set to allow detection of the gamma counts. Certain anatomic considerations will be helpful in SLN localization. If one were to draw a line along the lateral border of the pectoralis major muscle and lateral border of the latissimus dorsi muscle in the axilla, these would mark the outer borders of the axillary limits for the dissection. Careful marking of the axillary hairline is useful. One should place a tangential line at the axillary hairline in a perpendicular fashion anterior to posterior. A line is then drawn through the axis of the axilla, through the center point of the hairline. Those intersecting lines mark the center of a 5 cm circle, which can be drawn on the axilla. Within this 5 cm circle are 94% of the SLN (see Fig. 6.1). The remaining 6% will be found in the level II location. Using the gamma detection probe, this starting point of the 5 cm circle may be useful for identifying the actual location of the sentinel lymph node. Once the sentinel lymph node is found, an accurate incision may be made overlying the area of highest activity as determined by the gamma probe. The incision falls generally at or below the hairline as indicated above. Care should be taken to extend the dissection towards the chest perpendicular to the chest wall. The tendency of most novice surgeons is to make the incision and dissect in a cephalad direction. This should be avoided. Internally, the landmarks which localize the SLN are the central axillary vein and the third branch of the intercostal nerve. These anatomic structures are

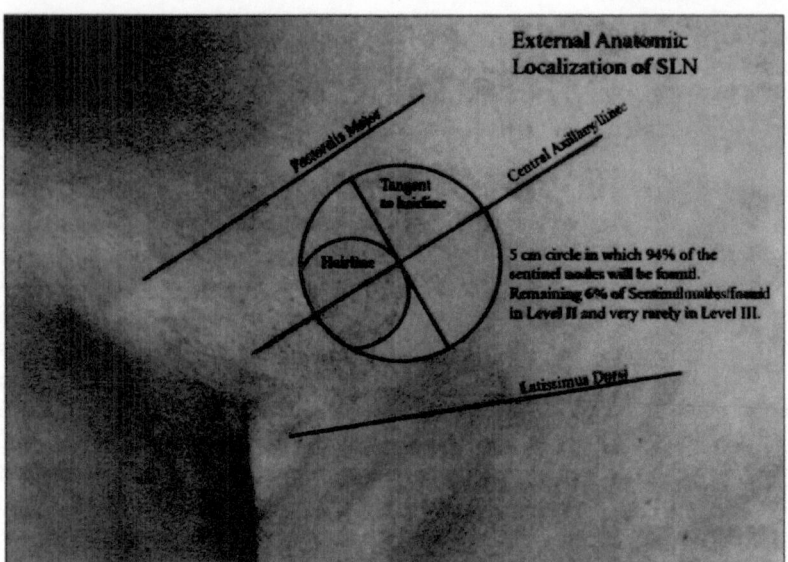

External Anatomic
Localization of SLN

Pectoralis Major

Central Axillary line

Tangent
to hairline

Hairline

5 cm circle in which 94% of the
sentinel nodes will be found.
Remaining 6% of Sentinel nodes/found
in Level II and very rarely in Level III.

Latissimus Dorsi

Fig. 6.1. (See Color Insert for color representation). Mapping the axillary SLN. If one were to draw a line along the lateral border of the pectoralis major muscle and lateral border of the latissimus dorsi muscle in the axilla, these would mark the outer borders of the axillary limits for the dissection. One should place a tangential line at the axillary hairline in a perpendicular fashion anterior to posterior. A line is then drawn through the axis of the axilla, through the center point of the hairline. Those intersecting lines mark the center of a 5 cm circle, which can be drawn on the axilla. Within this 5 cm circle are 94% of the SLN.

identifiable beneath the clavi-pectoral fascia. The central axillary vein can be found easily with careful dissection. Where the nerve crosses over the vein four quadrants are defined, as were seen on the external anatomy for the 5 cm circle. This internal anatomical point of intersection will define one of four quadrants, which collectively will contain 94% of the SLN, as noted above, found in breast lymphatic mapping.

GAMMA DETECTION

Intraoperative use of a gamma radiation detector can be complicated by several technical issues, including "shine through", and others, which are further discussed in another chapter.

"Shine Through" Effect

"Shine through" occurs when the gamma probe detects radiation energy from the injection site while looking for a SLN. It is generally due to a large deposit of radioactivity in a close proximity to the dissection area. It is critical to keep in mind exactly where the injection site is located in relation to where one is pointing the probe. An example of this is in a melanoma of the mid-back with an axillary SLN. Pointing the probe towards the axilla also aims it at the injection site in the back. Thus, one has to direct the aim of the probe towards the head or foot while searching the axilla. The same possibility exists with breast cancer in that the patient may have an ipsilateral lesion in the upper inner quadrant of the breast or bilateral breast cancers and have bilateral mapping performed with a contralateral lesion in the upper half of the breast. Each of these conditions may result in difficulty avoiding the "shine through" effect.

Another classic "shine through" problem is finding a hot node only to remove it and find that there is actually only minimal radioactivity in the removed node. This is usually due to another node containing the radioactivity directly behind the removed lymph node. Blue channels may surround the removed node, as well, but may actually be traveling directly to another node deep to the one removed. This reinforces the necessity of checking background counts after removal of the node to be sure all hot nodes were removed, and measuring the ex vivo counts of SLN removed to verify that indeed the node removed is hot. The hot nodes can sometimes be stacked so that even by removing the hot node another hot node lies directly behind. Again, this emphasizes the need to check the background as well as the node removed. Finally, another situation can make node detection with the probe difficult. Counts are drained through the blood vessels and occasionally the central axillary vein will be hot. This is verified by noting the course of the vein and measuring counts along the vein.

INTRAOPERATIVE MEASUREMENTS

There are several specific measurements that are important to validate the lymphatic mapping procedure. Included here will be information that is obtainable through the National Breast Lymphatic Mapping Database <http://surgonc.rad.usf.edu>. It is important to record the type of gamma probe used, the probe size, the time of injection and the dose of injection material. It is also important to measure the number of counts obtained at the incision site pre- and postexcision and the central axillary bed count.

The following data should be entered for each SLN harvested. By definition, SLN must be "hot", "blue" or "hot and blue". Hot nodes are defined as those with in vivo counts which are 2-3x those of the surrounding basin, or those with ex vivo counts which are 10x those of an excised nonsentinel node. Blue nodes are defined as those that are at least faintly stained by the isosulfan blue dye. A sentinel node in situ count should be obtained. The SLN should be exposed and the gamma probe should be placed directly on the exposed node before harvesting. The time of harvest should be entered as well. SLN ex vivo counts should also be recorded. For a measurement the node should be placed on a plain sterile towel located well away from the lymphatic basin and primary injection site. A postexcisional radioactivity count in the basin where the lymph node was removed should likewise be obtained. This measurement should be taken just after removal of the sentinel node at the site where the node once was. This count determines if the SLN has indeed been removed and detects the presence of additional SLN in the basin. Avoid directing the gamma probe toward the primary tumor site, which will result in "shine through" extraneous counts. Any non-SLN removed should have an ex vivo count recorded, in cps. Furthermore, a completion node dissection, if done, should be recorded and the reason for this procedure indicated. Reasons for complete lymph node dissection include: 1) node grossly positive (i.e., a node was grossly positive for metastatic breast cancer), 2) lymphatic mapping failed (i.e., lymphatic mapping failed to locate any hot or blue sentinel nodes), or 3) phase one patient (i.e., you are using a "phase one" protocol in which complete lymph node dissection is performed on all patients after SLN biopsy). Finally, the probe should be utilized to localize any counts extraneous to the injection site or axillary lymph node basin such as internal mammary nodes.

SPECIAL CASE: INTERNAL MAMMARY SLN

To date the incidence of internal mammary node involvement based on lymphatic mapping of the breast has been reported as 1-6% based on detected radioactive counts at the internal mammary location. Any central breast lesion or inner quadrant lesions of the breast should be evaluated with preoperative lymphoscintigraphy. Furthermore, when mapping any lesion of the breast it is critical to evaluate the internal mammary locations, specifically the second and third parasternal interspaces. Should "hot spots" be identified in these locations, it may be feasible to excise these internal mammary nodes. This has been accomplished by division of the intercostal muscles at the parasternal location thereby exposing the internal mammary nodes. These can be removed with careful dissection and application of the gamma probe to localize the node. This is one drawback to the blue dye method since it is nearly impossible to localize a node in the internal mammary location with the blue dye. To date, if a sentinel lymph node is located in the internal mammary area and cannot be removed easily because of the primary tumor location, especially in the event of a lumpectomy procedure, then we have postoperatively scanned the patient and tattooed the location of the internal mammary node in preparation for internal mammary lymph node radiation. If a mastectomy is being performed and the internal mammary lymph node is identified with the gamma probe, then the node is removed and submitted for histologic evaluation. If the internal mammary node returns positive on pathologic evaluation, then internal mammary radiation would be given.

SUMMARY

SLN mapping for breast cancer is technically feasible but there is a significant learning curve for all surgeons, even to an extent for surgeons experienced in SLN biopsy for melanoma. Most breast SLN are located in the axilla, rarely in the internal mammary chain. With appropriate gamma detection device selection and setup, attention to the technical details described here, and a little patience all surgeons should be able to learn this technique.

A national network of training centers is being established for radioguided surgery. This network will provide an opportunity for surgeons, nuclear medicine physicians and pathologists to come together and learn about this new technology. Training will include didactic sessions, live surgery, and hands-on experience with animal models. The faculty will consist of leading experts from acoss the country. Participating centers include the H. Lee Moffitt Cancer Center and Research Institute, John Wayne Cancer Institute, and the M.D. Anderson Cancer Center. Training sites will also be available in Durham, NC, Pittsburgh, PA, Seattle, WA, Little Rock, AR, and St. Louis, MO. In addition, a mentoring program will be available to assist the multidisciplinary lymphatic mapping team as they return to their institutions to implement this new procedure into their practices. Information on this national training and support network can be obtained by telephone (888-456-2840) and will soon be available at the web site <http://teleconmed.com>.

REFERENCES

1. U.S. Census Bureau, Population Division, Series P-25.
2. Kosary CL, Ries LAG, Hankey BF et al. SEER Cancer Statistics Review, 1973-1992: Tables and Graphs. National Cancer Institute. NIH Pub 1995; No. 95-2789.
3. Way LW. General surgery in evolution: Technology and competence. Am J Surg 1996; 171(1):2-9.
4. Geis WP, Kim HC, McAffe PC et al. Synergisitc benefits of combined technologies in complex, minimally invasive surgical procedures. Clinical experience and educational processes. Surg Endosc 1996; 10(10):1025-1028.
5. Cox CE, Pendas S, Cox JM et al. Guidelines for sentinel node biopsy and lymphatic mapping of patients with breast cancer. Ann Surg 1998; 227(5): 645-653.
6. Frazier TG, Copeland EM, Gallaher HS, Paulus DD Jr, White EC. Prognosis and treatment in minimal breast cancer. Amer J Surgery 1977; 133(6): 697-701.
7. Silverstein MJ, Rosser RJ, Gierson ED et al. Axillary lymph node dissection for intraductal carcinoma: Is it indicated? Cancer 1987; 59(10):1819-1824.
8. Balch CM, Singletary ES, Bland KI. Clinical decision-making in early breast cancer. Ann Surg 1993; 217:207-225.
9. Moffatt FL, Senofsky GM, Davis K et al. Axillary node dissection for early breast cancer: some is good but all is better. J Surg Oncol 1992; 51(1):8-13.
10. Silverstein MJ, Gierson ED, Waisman JR, Senofsky GM, Colburn WJ, Gamagami P. Axillary lymph node dissection for TIa breast carcinoma: is it indicated? Cancer 1994; 73(3):664-667.
11. Fisher B, Wolmak N, Bauer M et al. The accuracy of clinical nodal staging and of limited axillary dissection as a determinant of histologic nodal status in carcinoma of the breast. Surg Gynecol Obstet 1981; 152(6):765-772.
12. Cady B. The need to reexamine axillary lymph node dissection in invasive breast cancer. Cancer 1994; 73(3):505-508.
13. Morton DL, Wen DR, Wong JH et al. Technical details of intraoperative lymphatic mapping for early stage melanoma. Arch Surg 1992; 127(4):392-399.
14. Guiliano AE, Kirgan DM, Guenther JM, Morton DL. Lymphatic mapping and sentinel lymphadenectomy for breast cancer. Ann Surg 1994; 220:391-401.
15. Krag DN, Weaver DL, Alex JC et al. Surgical resection and radiolocalization of the sentinel lymph node in breast cancer using a gamma probe. Surg Oncol 1993; 2:335-339.
16. Albertini JJ, Lyman GH, Cox C et al. Lymphatic mapping and sentinel node biopsy in the patient with breast cancer. JAMA 1996; 276:1818-1822.

17. Borgstein PJ, Pijpers R, Comans EF et al. Sentinel lymph node biopsy in breast cancer: Guidelines and pitfalls of lymphoscintigraphy and gamma probe detection. J Amer Coll Surg 1998; 186(3):275-283.

18. Barnwell JM, Arredondo MA, Kollmorgen D et al. Sentinel node biopsy in breast cancer. Ann Surg Oncol 1998; 5(2:126-130.

19. Veronesi U, Paganelli G, Galimberti V et al. Can axillary dissection be avoided in breast cancer? Lancet 1997; 349:1864-1867.

20. Giuliano AE, Jones RC, Brennan M et al. Sentinel lymphadenectomy in breast cancer. J Clin Oncol 1997; 15:2345-2350.

21. Schreiber, SH, Pendas S, Ku NN, Reintgen DS, Shons AR, Berman C, Boulware D, Cox CE. Microstaging of breast cancer patients using cytokeratin staining of the sentinel lymph node. Ann Surg Oncol (in press: 1998).

22. Tiourina T, Arends B, Huysmans D, Rutten H, Lemaire B, Muller S. Evaluation of surgical gamma probes for radioguided sentinel node localization. European Journal of Nuclear Medicine (in press).

The Technique of Intraoperative Lymphatic Mapping and Sentinel Lymphadenectomy in Breast Cancer Using Blue Dye Alone

Philip I. Haigh, and Armando E. Giuliano

INTRODUCTION

The status of the axillary lymph nodes remains the most important factor associated with survival of patients with primary breast cancer. The standard method for staging is complete axillary lymph node dissection (ALND). Whether there is a survival benefit or local-regional benefit from ALND is controversial, with the debate far beyond the scope of this chapter. ALND to stage curable breast cancer patients will be performed in large numbers of patients *without* nodal disease and will subject them to the risks of nerve injury and lymphedema unnecessarily. The procedure using intraoperative lymphatic mapping and sentinel lymphadenectomy (SLND), as developed by Morton[1] to identify regional metastases from a primary cutaneous melanoma, was adapted by our group in an effort to identify axillary metastases in patients with breast cancer. The intent was to determine if the technique could be a viable alternative to routine ALND, accurately predicting the presence or absence of axillary metastases, without the associated morbidity of ALND. Using 1% isosulfan vital blue dye (Lymphazurin®) as the lymphatic mapping agent, 174 consecutive SLND procedures were performed in our initial feasibility study, followed by completion axillary lymph node dissection.[2] One goal was to establish the optimal technique necessary to identify the sentinel node. The sentinel node was identified in 114 of the 174 procedures (66%) and it accurately predicted the axillary nodal status in 109 (96%) cases. With the SLND procedure standardized, in a follow-up study using immunohistochemistry (IHC) stained on sections of the sentinel node, we found that SLND significantly increases the accuracy of detecting metastases in the axilla and also increases the

Radioguided Surgery, edited by Eric D. Whitman and Douglas Reintgen.
© 1999 Landes Bioscience

rate of detection when compared to ALND with routine histopathologic processing of random lymph nodes.[3] Furthermore, to validate the power of the sentinel node in predicting the entire axilla, 1087 nonsentinel nodes were examined in 60 patients whose sentinel nodes were tumor-free by both IHC and hematoxylin/eosin staining. Only one additional tumor-positive node was identified, indicating that the sentinel node is the most likely axillary node to contain metastases.[4] In our last report of 107 SLND procedures, using all refinements except preoperative lymphoscintigraphy for medial lesions, the technique was successful in identifying a sentinel node in 94% of cases, and was 100% predictive of axillary status.[5] Currently we identify the sentinel node in more that 99% of cases. The following is a description of this refined technique in detail, with key features of each step presented, which will hopefully serve to augment the accuracy of the procedure utilized by those surgeons who have been trained in the technique.

DETAILED TECHNIQUE

After induction of general anesthesia or deep IV sedation, the patient is prepared and draped with the arm free to allow for its movement. Under sterile conditions, 1% isosulfan blue vital dye is injected peritumorally into the breast parenchyma on the axillary side of the tumor (Fig. 7.1). It helps if the breast is elevated to prevent subfascial injection. If the patient has already had an excisional biopsy, then the dye is injected into the wall of the cavity and surrounding tissue. Biopsy cavity injection must be avoided, as lymphatic uptake will be minimal. If the patient had a core biopsy for a nonpalpable carcinoma, and is undergoing lumpectomy with pre-operative mammographic localization, then the injection is made at the appropriate depth and location around the wire, or if a needle is

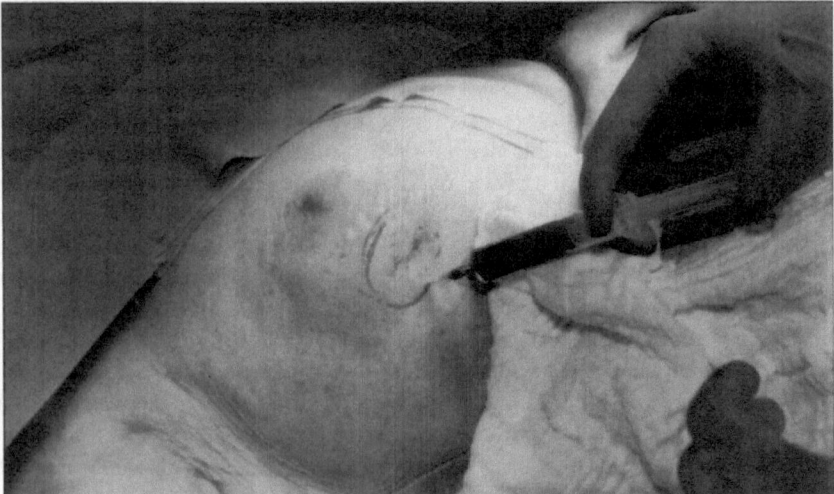

Fig. 7.1. (See Color Insert for color representation). Injection of isosulfan blue in the wall of the excisional biopsy cavity.

used then the dye may be injected using the same needle. The volume used ranges between 3-5 cc, with more dye used the further the lesion is from the axilla. The smaller volume is important for lesions high in the upper outer quadrant, because the axilla may be flooded with blue dye and make identification of the blue node particularly difficult.

After injection, massage or manual compression of the breast is essential to enhance lymphatic uptake of the dye. This compression is begun immediately after injection, continued for a specific time interval, and followed immediately by an axillary incision. The timing of the sentinel lymphadenectomy after dye injection depends on the location of the primary lesion. For high upper-outer quadrant lesions, the dissection should begin no sooner than three minustes after injection, whereas lower inner-quadrant lesions require seven to ten minutes for dye transit to the axilla. For all other areas, five minutes is the optimal time for sentinel lymph node dissection. This timing is crucial, for too short an elapsed time does not allow dye to reach the sentinel node, while a delay may stain nodes blue "beyond" just the sentinel node, or allow egress of dye from the sentinel node making identification troublesome.

Sentinel lymphadenectomy is performed using a small transverse incision about 1 cm inferior to the hair-bearing area of the axilla, and slightly towards the anterior axillary line (Fig. 7.2). The subcutaneous tissue is incised with electrocautery, and dissection is continued directly perpendicular to the skin. There are often superficial subcutaneous blue-stained lymphatics, which can be transected; these should not be confused with lymphatics that drain the primary tumor. At this point, the shoulder should be abducted and flexed which often helps to bring the axillary contents anteriorly closer to the incision. With meticulous dissection using

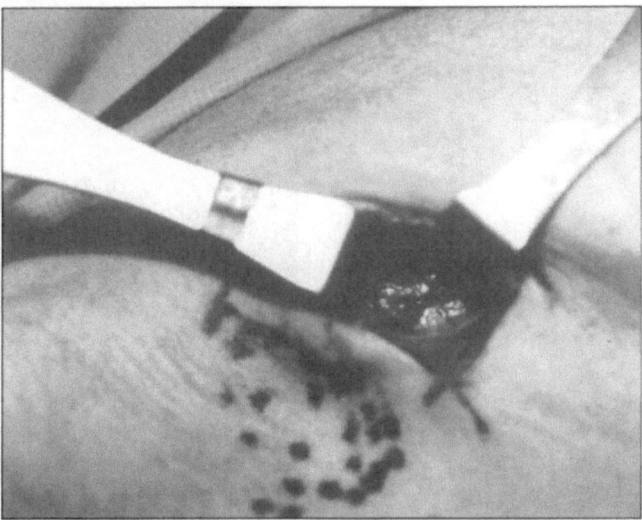

Fig. 7.2. (See Color Insert for color representation). Axillary incision about 1 cm below hair bearing area (indicated with purple markings).

fine Crile forceps, a blue lymphatic can be identified and easily traced to a blue lymph node (Figs. 7.3a,b). The field must be kept dry as any bleeding rapidly obscures the contrast of blue dye to the yellow color of fat. If a blue lymphatic is identified, carefully follow this superiorly and inferiorly by incising surrounding tissue and clavi-pectoral fascia to identify a blue node. This method simplifies

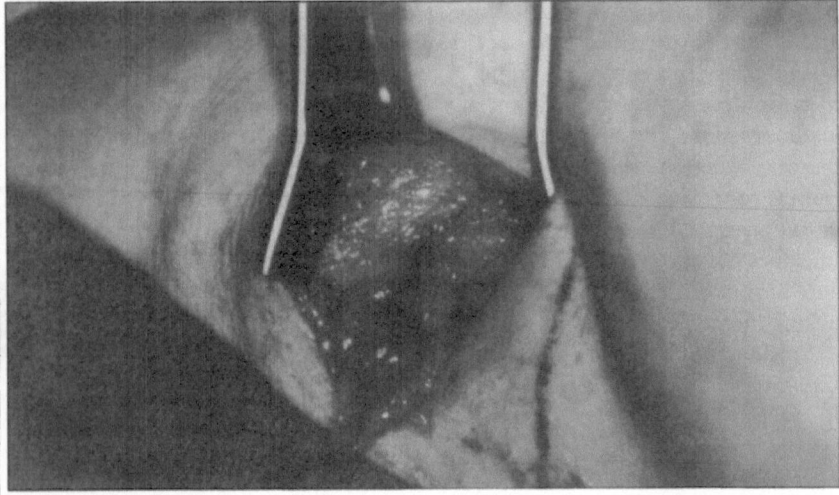

Fig. 7.3a. (See Color Insert for color representation). Blue lymphatic entering into blue-stained sentinel lymph node.

Fig. 7.3b. (See Color Insert for color representation). Another example of a blue sentinel lymph node in vivo.

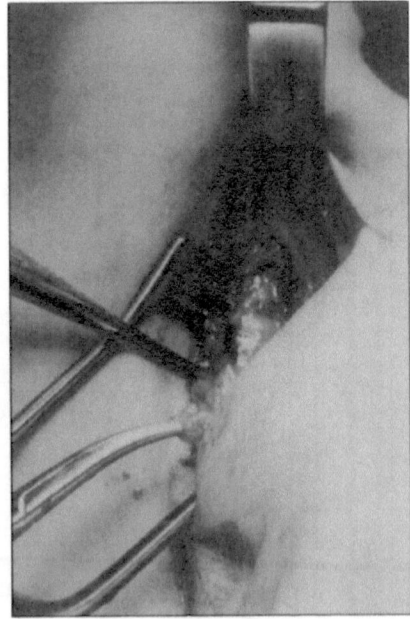

identification of a blue node, which may only have a small portion of its hilum stained blue near the afferent lymphatic. If a blue lymphatic channel is not identified, blindly searching for a blue node is tedious, difficult, and often not successful. Similarly, the blue lymphatic must be kept intact to facilitate identification of the blue node. After excising the blue node, it is imperative that the afferent blue lymphatic is followed towards the breast to ensure there is not another blue lymph node more proximally representing the true sentinel lymph node (Fig. 7.4).

The excised sentinel node(s) is then bisected and frozen section obtained. If negative, then the wound is irrigated and closed without drainage. According to our current research protocol, a completion axillary lymph node dissection is not performed if the sentinel lymph node is free of metastases. If frozen sections are postive for metastases, then complete axillary lymph node dissection is performed immediately. The sentinel lymph node is processed for permanent section evaluation by hematoxylin and eosin staining, and if negative, further levels of the bisected node are stained for cytokeratin using immunohistochemistry. If metastases are identified after permanent sections are reviewed, then the completion lymphadenectomy is performed within a week after SLND. After sentinel lymphadenectomy, the primary lesion or biopsy cavity is removed with standard breast conservation surgery or total mastectomy. The significance of IHC-detected tumor deposits is however unknown and therefore should not be a standard procedure outside of a research protocol.

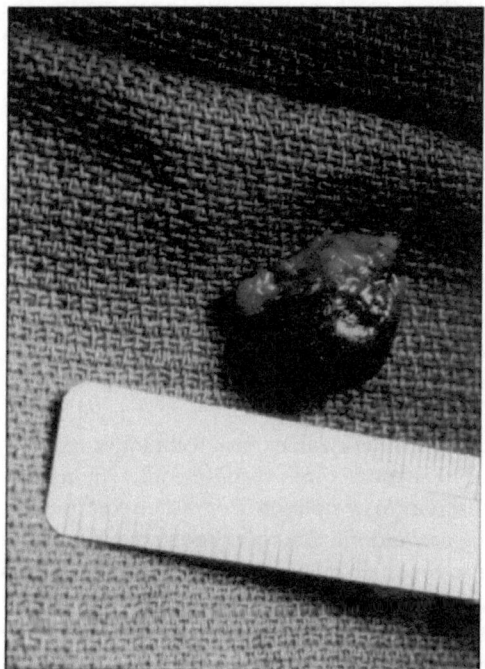

Fig. 7.4. (See Color Insert for color representation). Excised blue sentinel lymph node.

COMPLICATIONS

Complications of this surgical procedure can occur from either the isosulfan blue dye, or the surgery itself. In our experience, there have been no injuries to the long thoracic or thoracodorsal nerves as a result of sentinel node biopsy. Other complications, including infection, lymphocele, or other local or regional problems have occurred rarely, as they might from any axillary lymph node biopsy procedure, without any apparent increase associated with sentinel node surgery specifically.

Side effects from the blue dye may occur. After intra-parenchymal breast injection the dye typically migrates to the dermal lymphatics, causing the skin overlying the injection site to become blue-stained (although less so than with intradermal injection). This generally fades rather quickly, but the occasional patient may have residual faint bluish discoloration of the breast skin for months. This almost never occurs following total mastectomy or re-excision of the breast cancer at the same time as sentinel node biopsy, but is more likely when the sentinel node biopsy is performed *after* breast lumpectomy. All patients should be informed that their urine will turn a green color for 24-48 hours after the procedure, a predictable but self-limited process as residual dye is cleared from the patient's body.

Other complications of blue dye injection have been reported in up to 1.5% of patients.[6] Most of these reactions were of an allergic type, consisting of localized swelling at the injection site and mild pruritus of the hands, abdomen, and neck within minutes of injection. More rarely, more severe reactions have been reported,[7,8] although at least some of these may have been due to mixing the dye with local anesthetic, which is known to cause a precipitate. These have consisted of facial and glottic edema which has progressed in extremely rare cases to respiratory distress or shock, with a reported fatality in a related compound used to estimate the depth of a burn.[6]

The blue dye may also interfere with the function of percutaneous oximetry probes. Chemically related dyes, (i.e., Patent Blue V), have been shown to absorb light at a wavelength similar to deoxygenated hemoglobin, resulting in diminished oxygen saturation readings by standard oximetry probes. Technically, the patient's blood absorbance of red light is increased by tri-aryl methane dyes (such as Patent Blue V or Lymphazurin) relative to its absorbance of infra-red light. As this ratio increases, the calibrated (calculated) oxygen saturation linearly decreases. This appears more likely in anemic patients and at higher intravascular dye concentrations.[8,9]

Overall, the incidence of these complications with current techniques and dosages is unknown but appears low. The manufacturer[6] recommends avoiding usage in patients with known hypersensitivity to Lymphazurin or related compounds, pregnant women, nursing mothers, or children, due to unknown risks in those latter three groups. The carcinogenic potential of Lymphazurin is unknown, as is its effect, if any, on male or female fertility. Lymphazurin should never be mixed with local anesthetics because a precipitate immediately forms. Finally, due to the side effects described above, administration and observation by trained/informed medical personnel, and the availability of emergency medical facilities, is recommended for 30-60 minutes after administration, a requirement easily met in the clinical context of sentinel node mapping as currently practiced at our institution and others.

SUMMARY

The SLND technique using blue dye alone as developed over the last seven years at our institution has been proven to precisely predict the status of the axillary lymph nodes. The procedure is well tolerated, and axillary staging can be achieved with minimal morbidity. At first the procedure can be annoyingly frustrating, but as with any operation, patience, keen attention to detail, and adherence to sound surgical principles will lead to success. The detailed description of the procedure and technical hints will hopefully aid surgeons who have adopted the procedure into their own practice in identifying the sentinel node with confidence. Surgeons should validate the technique at their own insitutions to ensure accuracy; mastery in removing the true sentinel node will not help if the pathologists cannot detect metastases with proficiency.

REFERENCES

1.   Morton DL, Wen D-R, Wong JH et al. Technical details of intraoperative lymphatic mapping for early stage melanoma. Arch Surg 1992; 127:392-399.
2.   Giuliano AE, Kirgan DM, Guenther JM, Morton DM. Lymphatic mapping and sentinel lymphadenectomy for breast cancer. Ann Surg 1994; 220:391-401.
3.   Giuliano AE, Dale PS, Turner RT et al. Improved axillary staging of breast cancer with sentinel lympyhadenectomy. Ann Surg 1995; 222:394-401.
4.   Turner RT, Ollilla DW, Krasne DL et al. Histopathologic validation of the sentinel lymph node hypothesis for breast carcinoma. Ann Surg 1997; 226(3):271-278.
5.   Biuliano AE, Joens RC, Brennan M, Statman R. Sentinel lymphadenectomy in breast cancer. J Clin Onc 1997; 15(6):2345-2350.
6.   Product Insert: Lymphazurin 1% (isosulfan blue). Norwalk, CT: United States Surgical Corporation, 1997.
7.   Hietala SO, Hirsch JI, Faunce HF. Allergic reaction to patent blue violet during lymphography. Lymphology 1977;10:158-160.
8.   Longnecker SM, Guzzardo MM, Van Voris LP. Life-threatening anaphylaxis following subcutaneous administration of isosulfan blue 1%. Clin Pharm 1985;4:219-221.
9.   Larsen VH, Freudendal-Pedersen A, Fogh-Andersen N. The influence of Patent Blue V on pulse oximetry and haemoximetry. Acta Anaesthesiol Scand 1995;39 Suppl 107:53-55.
10. Saito S, Fukura H, Shimada H, Fujita T. Prolonged interference of blue dye "patent blue" with pulse oximetry readings. Acta Anaesthesiol Scand 1995;39:268-269.

Lymphoscintigraphy

Claudia G. Berman

INTRODUCTION

A diagnostic strategy utilizing selective lymph node sampling directed by lymphoscintigraphic methods has been advocated in recent years. It has been proposed that meticulous pathological study of the sentinel lymph node (SLN), the primary lymph node draining the malignant lesion, will accurately predict the pathologic status of the remainder of the draining lymph node basin. This hypothesis has been shown to be true, in both melanoma and more recently carcinoma of the breast.[1] Not only nodes of the draining basin but in-transit lymph nodes, situated between the injection site and the anatomically recognized regional lymph node groups, have been found to accurately predict the pathologic status of the regional nodal basin as a whole (Fig. 8.1). The prospect of less morbid and more accurate tumor staging is apparent and appealing.[2]

Lymphoscintigraphy is the injection of radioactive particles around a tumor in an effort to identify the lymph nodes that have afferent drainage from that tumor. Lymphoscintigraphy is usually performed using technetium, which is an ideal radionuclide for imaging. All equipment in modern nuclear medicine departments is optimized for the technetium energy peak. Technetium provides a low radiation dose to the patient(s) and staff. It is a gamma emitter only, unlike other agents such as iodine-131 and gold, which are beta emitters. Technetium has a short half-life of approximately 6 hours. Again, this means that the radiation dose for the patient and staff members is lessened, and it also gives the ability to perform repeat studies over a short period of time if this is desirable or necessary.

The choice of the particle to which the technetium is to be labeled varies with each diagnostic problem. It is by no means certain what is the optimal particle for lymphoscintigraphy. If a particle is too small, it will be taken up preferentially into the capillaries which results in less of the dose going to the lymph nodes and a low target to background ratio. If a particle is too large, it will not be taken up into lymphatics but will be phagocytized. This results in less uptake in the lymph nodes and more uptake at the injection site.

Radioguided Surgery, edited by Eric D. Whitman and Douglas Reintgen.
© 1999 Landes Bioscience

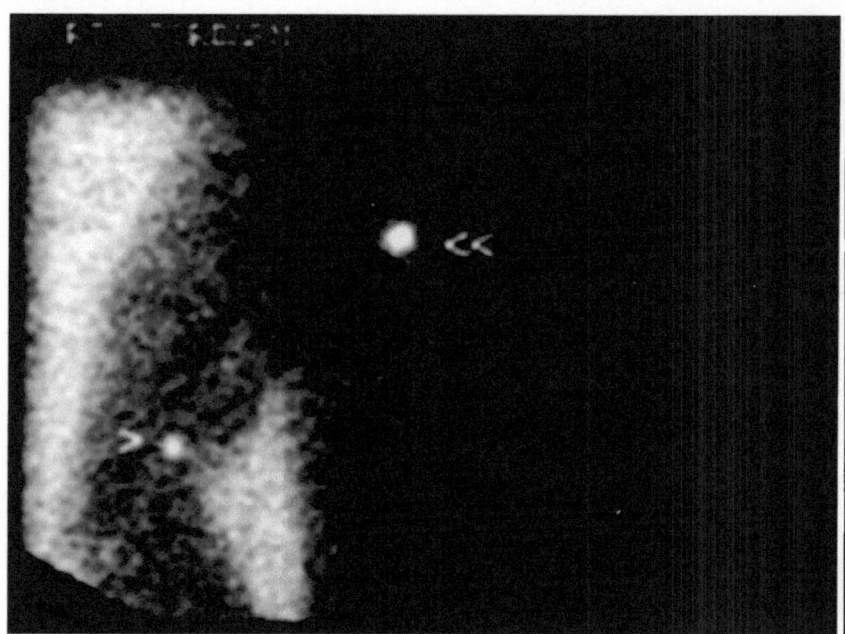

Fig. 8.1. Anterior view, right arm and axilla, shows uptake in the antecubital fossa (<) which represents an in-transit lymph node which by definition is a SLN. There is also uptake in the axilla (<<).

Human serum albumin is commonly used for lymphoscintigraphy because of its uniform particle size. I do not advocate the use of albumin because it seems to pass through the lymph nodes rapidly. Instead of uptake in one or two SLN, we commonly see six to eight lymph nodes with radioactive uptake following the injection of radiolabelled albumin. We also worried that since the albumin passes through the lymph nodes so rapidly, that it may in fact pass through the SLN by the time the patient is presented to the operating room, thus defeating the purpose of SLN mapping. It is also more difficult to use the hand held gamma probe and identify SLN when there are a number of lymph nodes which are radioactive (Fig. 8.2).

Historically, investigators have preferred technetium antimony trisulfide colloid for lymphoscintigraphic studies because its particle size, 3-30 nm, is optimal for transit through lymphatics and localization in nodes without phagocytosis. Unfortunately, antimony sulfur colloid is no longer produced for use in the United States. It is still available in Australia and Europe and is used widely there. Hung and associates have produced acceptable lymphoscintigraphic images with technetium sulfur colloid filtered to a maximum particle size of 100 nm. In practice, nearly all of their filtered activity was distributed in the 15-50 nm range.[2]

We too use sulfur colloid. Our filtration technique uses a 0.2 micron (20 nm) filter and we have found this preparation satisfactory. It is important that the sulfur colloid be fresh because over time there is clumping of the colloidal particles,

Fig. 8.2. Left anterior oblique head and neck lymphoscintigram; (A) immediately after, (B) 15 minutes and (C) 45 minutes post-injection. The immediate images show uptake in a SLN (closest to injection site). The lymphatics are bifurcated and two non-SLNs are also seen. After 15 minutes, additional non-SLNs are present and after 45 minutes many non-SLNs are present making the detection of the SLN much more difficult.

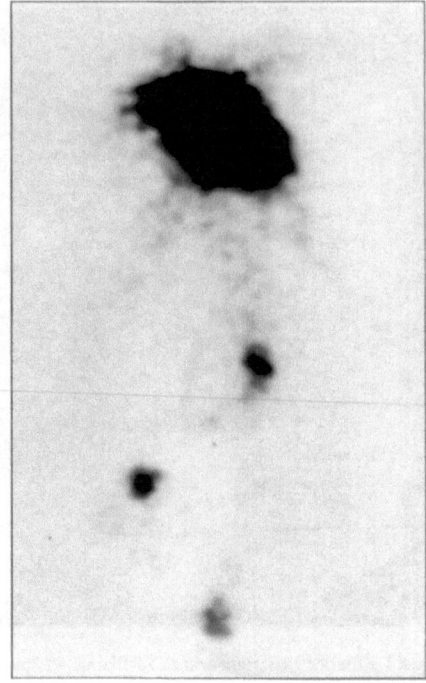

Fig. 8.2.A

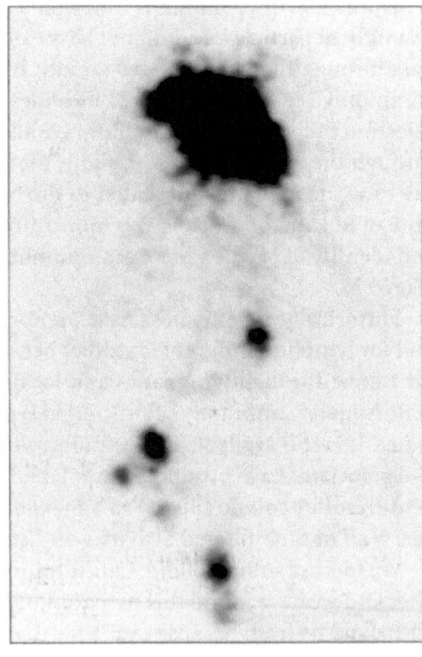

Fig. 8.2.B

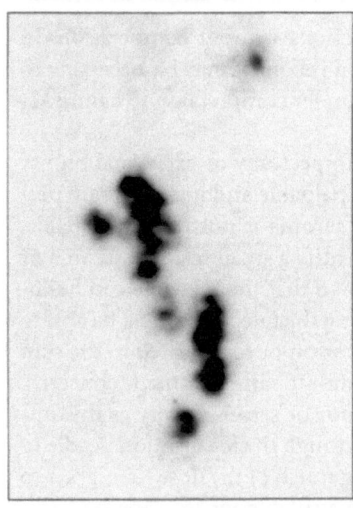

Fig. 8.2.C

which in effect causes them to be larger in size and less effective. Our 0.2 micron sulfur colloid preparation appears to work very reliably, consistently identifying one or two SLN for excision.

The equipment necessary for doing lymphoscintigraphy is commonly found in most nuclear medicine departments. We use a standard large field of view gamma camera, a high resolution collimator and a 10% imaging window at the 140 keV energy peak, which is the technetium energy peak.

In most situations we employ a dose of 450 microcuries. This includes cases involving the breasts, trunk and extremities. If the lesion is very near a draining lymph node basin, we may use 250 microcuries to minimize artifact from the injection site. We do this only rarely, for example in breast cases where the lesion is well within the axilla. I also advocate using 250 microcuries in head and neck cases. The anatomy of the head and neck is very tight and compact and it is often difficult to identify SLN because of interference produced by the injection site. Minimizing both the volume and dose injected seems to improve results.

DOSE ADMINISTRATION

BREAST
Patients with palpable tumors are injected in the nuclear medicine department with six injections around the periphery of the tumor at the depth of the mass. Patients with nonpalpable tumors first undergo mammographic or ultrasound needle localization. If imaged by ultrasound six equal aliquots are injected, with ultrasound guidance if necessary, at the correct depth. If imaged by mammography six equal aliquots are injected equidistant from the tip of the localization wire, at the correct depth. In small breasted patients or patients with superficial

tumors, it is often possible to perform the injections without compression. In large patients or in patients with centrally located lesions it may be necessary to perform the injections while the breast remains under compression to assure accurate depth.

A third group of patients are studied after lumpectomy or excisional biopsy has been performed. Most have seromas that are palpable and injections are performed as if the seroma were the tumor. If the seroma is not readily palpable, ultrasound is used and in almost all cases will identify a small residual seroma or area of architectural distortion within the breast so that the injectate can be instilled at the correct location. The physician making the injections must be able to distinguish the tumor seroma or scar from induration or scar related to the skin incision, which is often placed remote from the tumor to satisfy cosmetic concerns.

Care must be taken not to inject into the tumor or seroma cavity as this impedes lymphatic flow. We have found injection through the localization needle to be undesirable as the needle acts as a wick causing much of the dose to migrate to the skin surface. This not only reduces the dose available for imaging but produces confounding contamination on the skin surface (Fig. 8.3).

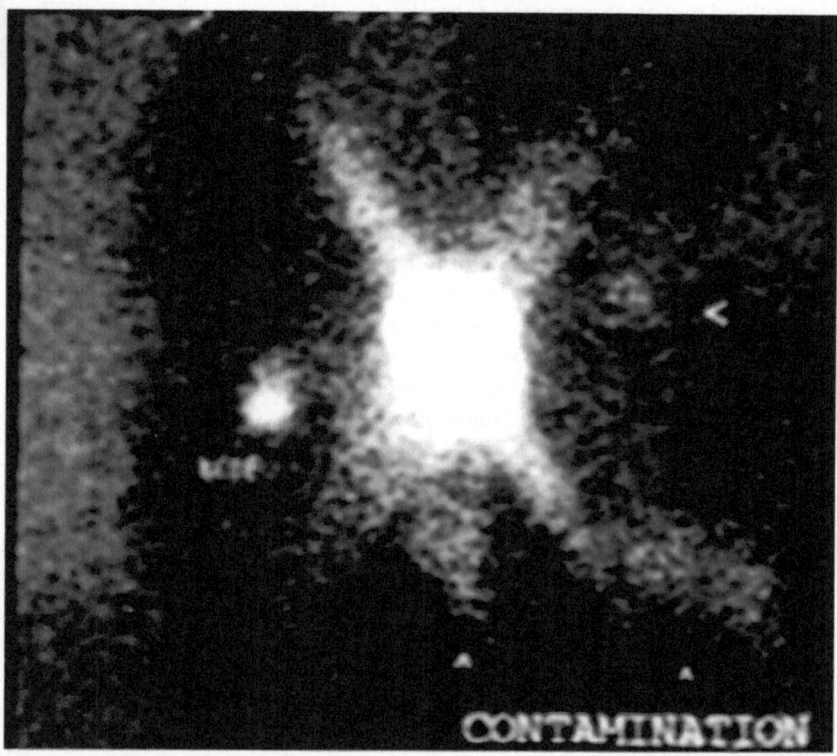

Fig. 8.3. Breast lymphoscintigraphy, right anterior oblique projection. This right breast carcinoma shows axillary drainage. The more diffuse areas of uptake (<) are the result of contamination from the "wick effect" of a localization needle.

We have obtained good results using both 2 cc and 6 cc total diluent. The ideal volume of injectate is unknown. On one hand a larger volume should improve uptake within the lymphatic system. On the other hand larger volumes may diffuse into the axilla, confounding imaging and detection. We are currently performing a randomized study comparing results between injected volumes of 2 cc versus 6 cc.

MELANOMA

When patients present with an intact melanoma, the radiopharmaceutical is injected about the periphery of the melanoma in four equal aliquots. One cc of total volume diluent is used. The injections are performed intradermally and care is taken to obtain a skin wheal. Many cases will present following excision (biopsy or wide local excision) of the primary melanoma. The four injections are performed about the center of the scar. Two injections are placed on each side of the scar. The injections are approximately 1 centimeter apart. It is important that the injections not be performed at the ends of the scar because drainage from the ends may not necessarily reflect the drainage pattern of the original tumor.

IMAGING PROTOCOL

A caution generally pertinent to the use of lymphoscintigraphy for SLN identification relates to the importance of confirming that the lymph nodes that pick up the radiopharmaceutical are indeed SLN. The nuclear physician must track the afferent drainage channels to see if multiple drainage channels culminate in multiple SLN (Fig. 8.4) or if the lymphatic channels each terminate in a single SLN (Fig. 8.5). One must also ascertain whether or not the visualized lymph nodes run in series, meaning that the first lymph node is the sentinel node, or in parallel, meaning that each imaged lymph node represents a SLN.

BREAST

The breast carcinoma patient is imaged immediately after injection, positioned supine under the gamma camera in the anterior oblique lateral projection. The arm is extended above the head and the hand placed under the head to optimize axillary exposure. Unlike imaging in melanoma, it is very unusual to see afferent lymphatics. As the regions of interest are the axilla, clavicular region and internal mammary nodes, various maneuvers are attempted to remove breast activity from the camera's field of view. The injection site in the breast can be shielded with lead, but this may produce a penumbra effect which can camouflage lymph nodes. The breast can sometimes be taped out of the field of view. Alternatively, the patient may be imaged sitting or standing so that gravity will act to remove the injection site from the field of view. Small breasted women or women with lesions near the chest wall or axilla often present a challenge and it may not be possible to separate the injection site from the regional lymph node groups.

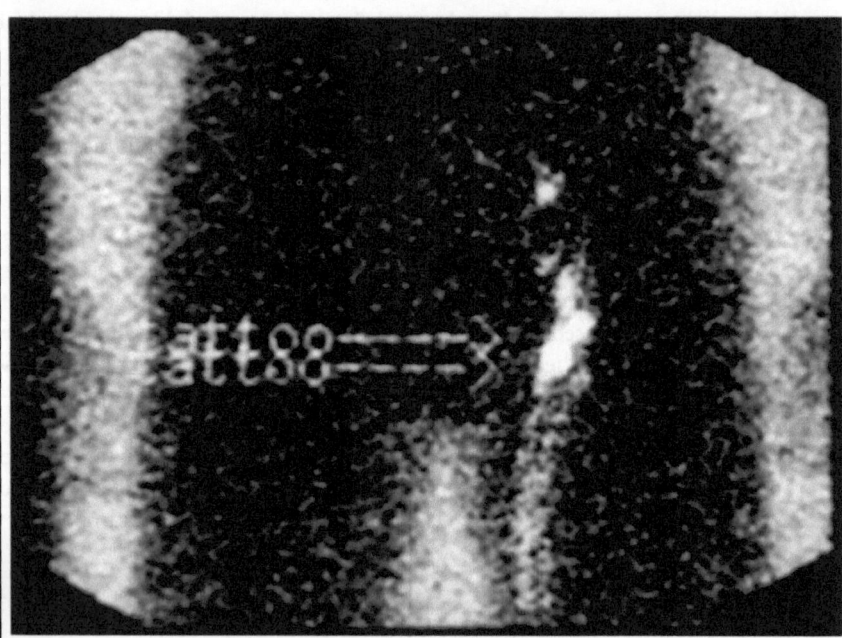

Fig. 8.4. Anterior pelvic lymphoscintigraphy shows two afferent lymphatics which each end in a lymph node, resulting in two SLNs (>).

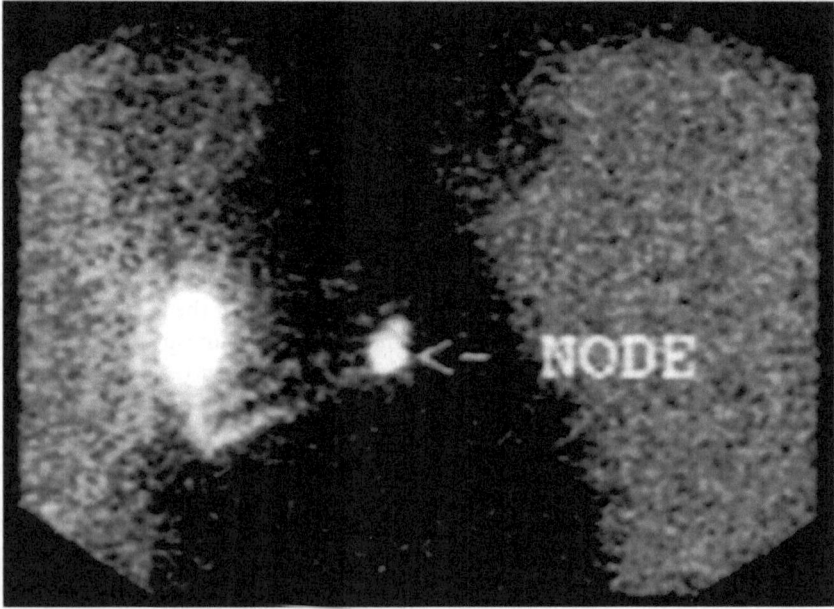

Fig. 8.5. Lateral chest lymphoscintigraphy shows two afferent lymphatics ending in a single axillary lymph node (<).

The persistence scope is used to identify accumulations of the radiopharmaceutical corresponding to lymph nodes. Internal mammary and supraclavicular lymph nodes can be tattooed in the anterior projection, whilst the axillary lymph nodes are tattooed with the patient in the lateral position with the arm above the head. Images are acquired over 8-10 minutes per view to assure high count density. Images are obtained in the anterior, lateral and oblique positions. The patient can be imaged on a cobalt flood source to define the body contour. A cobalt flood source is a sheet of radioactive material that is giving off gamma rays. The patient lies on the flood source and is between the flood source and the camera. Therefore, the patient's body will attenuate gamma rays and we are able to see the body contour (Fig. 8.6). Alternatively, the body contour and landmarks can be demonstrated using a Technetium-99m marker. SLN are found with immediate imaging in approximately 60% of cases. Delayed imaging will be necessary in the remainder of cases. In our practice the optimal time for delayed imaging as well as for surgery with hand held gamma probe guidance is 4-6 hours following injection.

Conventional wisdom, noting that most breast carcinomas arise within the upper outer quadrant of the breast, holds that nodal drainage will be to the axilla and only the most medial tumors will drain to the internal mammary chain. Uren studied 32 patients with antimony sulfur colloid lymphoscintigraphy and found that there was ipsilateral axillary node drainage in 85% of the cases.[3] However the

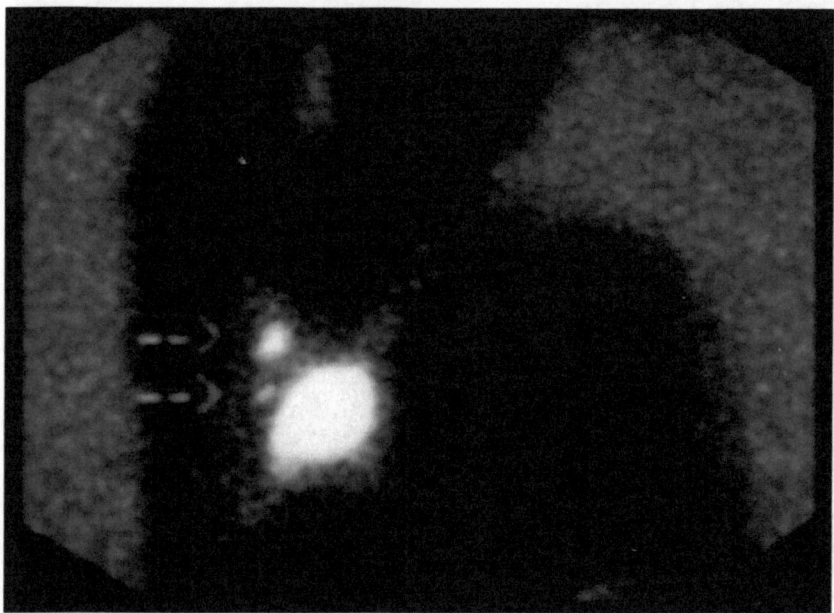

Fig. 8.6. Right anterior oblique view, breast lymphoscintigram, showing uptake in two lymph nodes in the right axilla. The patient is imaged on a cobalt flood source.

multiplicity and variability of drainage patterns was unexpected. Twenty-eight percent of patients with outer quadrant tumors showed unexpected drainage to internal mammary nodes while 33% of inner quadrant tumors showed axillary drainage. Thus one-third of patients with lateralized tumors had drainage which crossed the midline of the breast. Twenty percent of patients with upper quadrant tumors showed direct drainage to supraclavicular or infraclavicular nodes. In one patient an in-transit intramammary node, lying in the breast parenchyma between the primary lesion and the axilla, was discovered. This in-transit lymph node was in fact the SLN and contained metastatic disease (Fig. 8.7). Standard axillary dissection would not have identified this node and, by implication, the patient's need for systemic adjuvant therapy.

Preoperative breast lymphoscintigraphy offers the opportunity for identification of the unique pattern of nodal drainage for each malignant lesion. Directing the surgeon to the site of the SLN minimizes the operative time, the extent of dissection and likelihood of late morbidity. Identification of in-transit lymph nodes

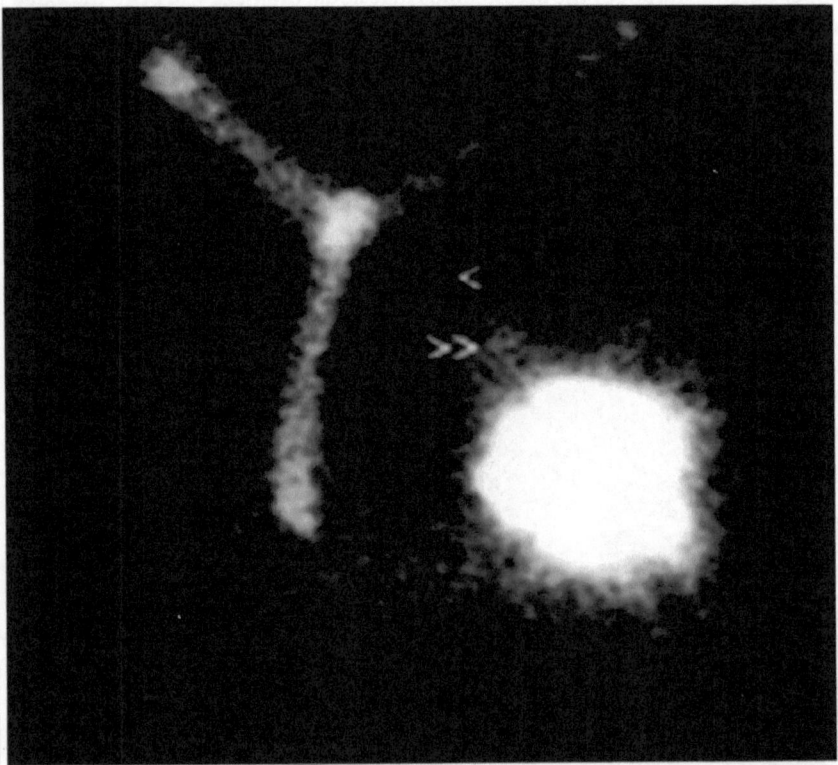

Fig. 8.7. Breast lymphoscintigraphy, anterior projection, clavicles and sternum are outlined by a Tc-99m marker. This periareolar cancer shows flow to an intramammary lymph node (<<), and a left internal mammary lymph node (<). Since afferent lymphatics are not identified, it is not possible to determine whether one or both lymph nodes are sentinel. Both were marked for excision.

opens up a subset of patients inadequately examined by standard axillary dissection techniques. Identification of internal mammary SLN allows for rational radiotherapeutic planning in patients who might not otherwise be recognized to be at high risk (Fig. 8.8). The identification of a supraclavicular lymph node as a SLN is of unknown significance given the current American Joint Commission on Cancer (AJCC) staging system which classifies supraclavicular adenopathy as metastatic, stage IV, breast carcinoma.

MELANOMA

Melanoma, in distinction to carcinoma of the breast, is a much easier disease in which to successfully perform lymphoscintigraphy due to the richness of the cutaneous lymphatics. The patient is placed under the gamma camera immediately following injection. An effort is made to identify the afferent lymphatics. We feel, as do others, that immediate ("dynamic") scanning is an essential component of any SLN identification procedure (Fig. 8.9).[4] The afferent lymphatics are followed with the gamma camera using the persistence scope. The SLN and intransit lymph nodes are localized, moving the patient in multiple projections, and they are marked or tattooed. Multiple drainage basins may require additional imaging. If the multiple drainage basins are in different locations of the body it may be necessary for the patient to have more than one imaging session. Delayed imaging is seldom required.

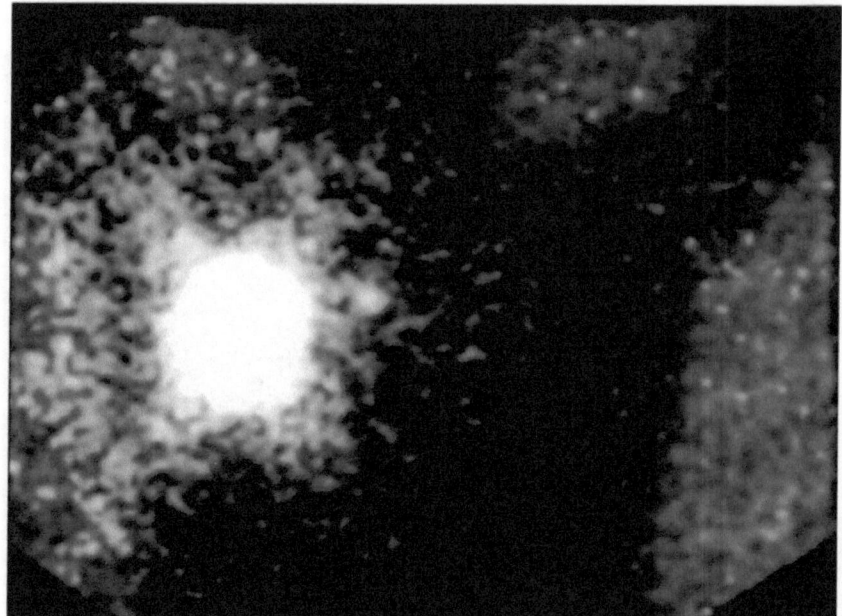

Fig. 8.8. Anterior view, breast lymphoscintigraphy, showing flow to right internal mammary chain only in a lesion at about the 12 o'clock position of the right breast.

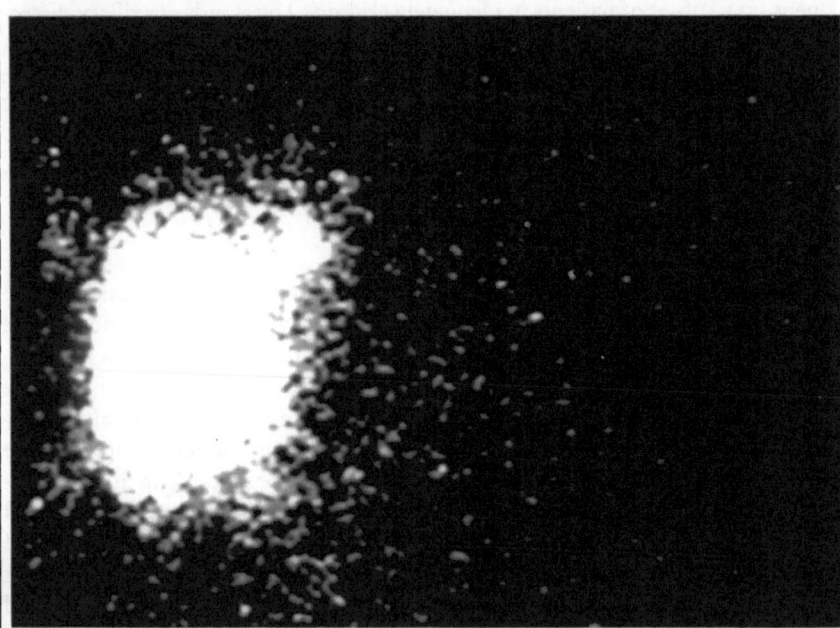

Fig. 8.9. Anterior chest lymphoscintigram in a patient with melanoma of the right arm. (A) immediate view-shows drainage to a single lymph node.

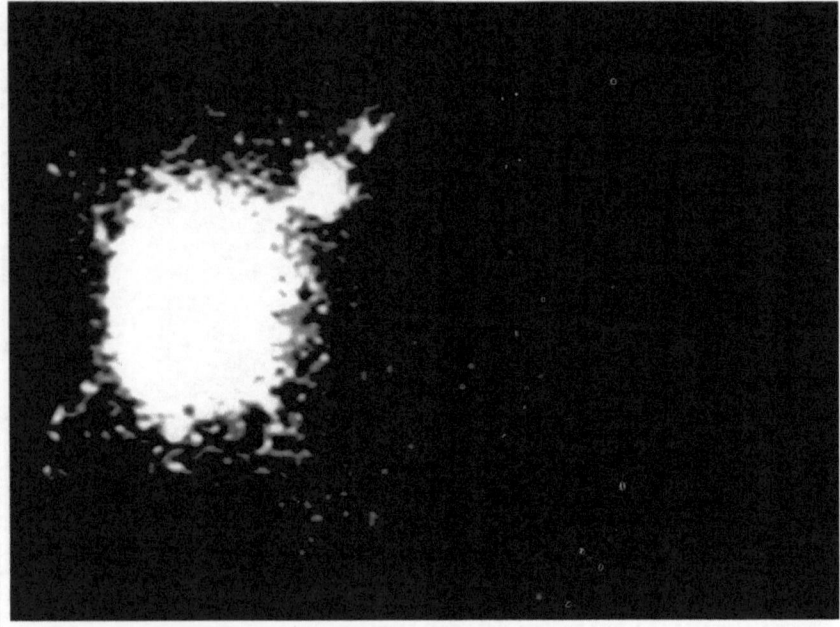

Fig. 8.9. (B) ten minute delay-shows uptake in a second lymph node.

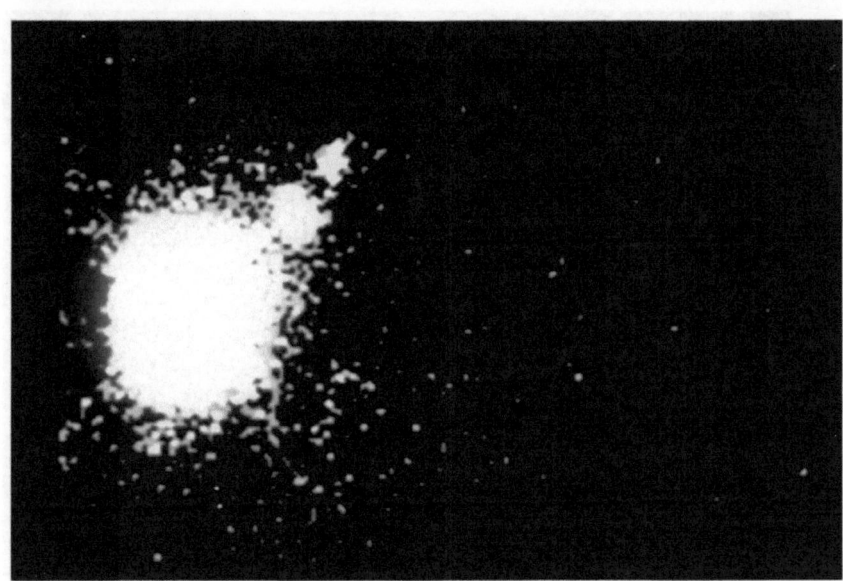

Fig. 8.9. (C) fifteen minute delay-shows continuing increase in uptake in both nodes.

8

Localization of the lymph nodes is performed by positioning the patient in various projections and using the persistence scope and hot marker to triangulate upon the lymph nodes (Fig. 8.10). We routinely outline the body contour and body landmarks, such as the sternocleidomastoid and ear, with a hot technetium marker (Fig. 8.11). The patients are imaged on a cobalt flood source to show the body contour.

RADIOGUIDED SURGERY

Considerable recent interest has centered around a strategy of injecting targeted radioisotopes systemically. This is to be followed by surgical exploration using a radioactivity-sensitive hand-held probe to guide the surgeon to otherwise indiscernible tumor deposits or tissues. Two distinct targeting strategies readily present themselves. For tumor, the most attractive approach in most settings would employ a radioactively labeled antibody specific for a definitive tumor antigen. For metabolically distinctive tissues a radioactively labeled precursor molecule would represent an 'ideal' targeting agent.

Ongoing efforts using various radioactively-labeled preparations in a wide array of tumors and clinical situations remain in a developmental stage. At the present time only a few such techniques are used regularly in our department, recommended by consistent clinical reliability.

We have used Tc-99m-oxydronate, one of several standard bone imaging agents, to localize bone tumors for excisional biopsies. The patient is injected with the

Fig. 8.10. (A) Anterior, (B) left and (C) right lateral chest lymphoscintigrams showing separate drainage into both axillae. The body contour is outlined by placing the patient on a cobalt flood source.

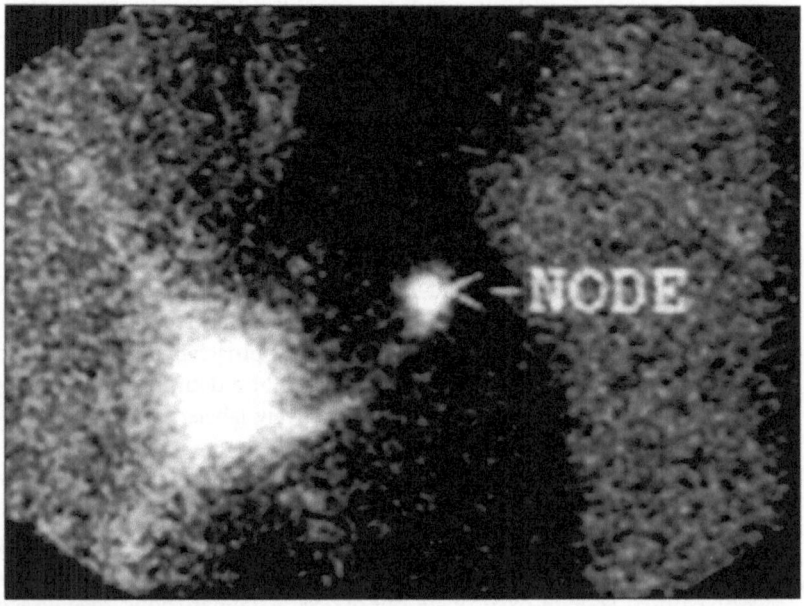

Fig. 8.10. (B)

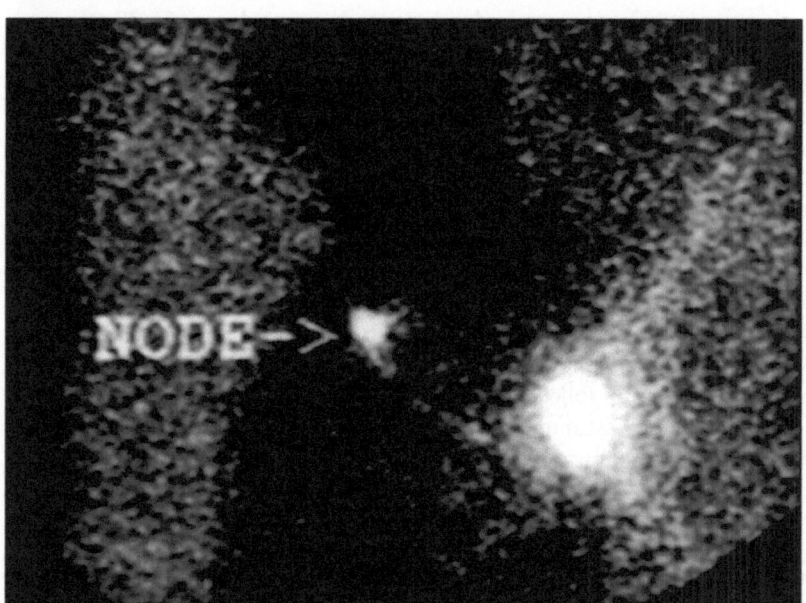

Fig. 8.10. (C)

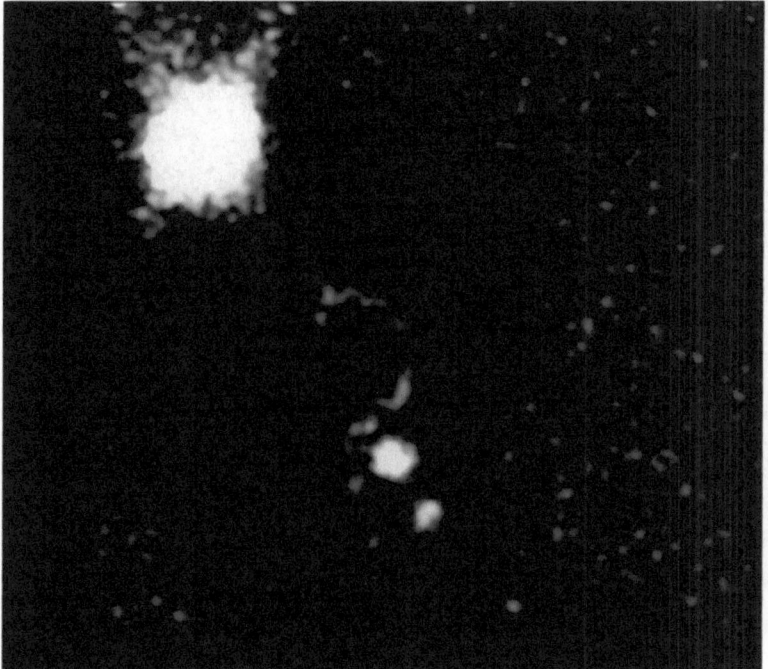

Fig. 8.11. Left lateral head and neck lymphoscintigram showing drainage into the posterior cervical chain only in a patient with a midline forehead melanoma. The ear and sternocleidomastoid are outlined with a Tc-99m marker.

radiopharmaceutical using standard techniques. We use a dose of 25-30 mCi. and routinely use intravenous hydration to increase wash out of soft tissue uptake. Unlike standard bone imaging protocols, however, the injection is timed to precede surgery by 5-6 hours in an effort to optimize target to background ratios. Imaging has invariably been performed previously and is not repeated at the time of surgery aside from imaging of the specimen to verify removal of the targeted abnormality.

We also routinely use Tc-99m-Sestamibi to aid the surgeon performing parathyroidectomy in patients with hyperparathyroidism from parathyroid adenoma. Sestamibi is preferentially taken up in the mitochondria of tissues undergoing oxidative metabolism. Uptake is pronounced in both the thyroid and parathyroid. However, washout is much more rapid from thyroid tissue. The patient is injected with 25 mCi. Sestamibi followed at one hour by imaging of the neck and upper two-thirds of the mediastinum. The patient is positioned supine and the head and neck are extended and immobilized. Images are acquired in the anterior and 30%

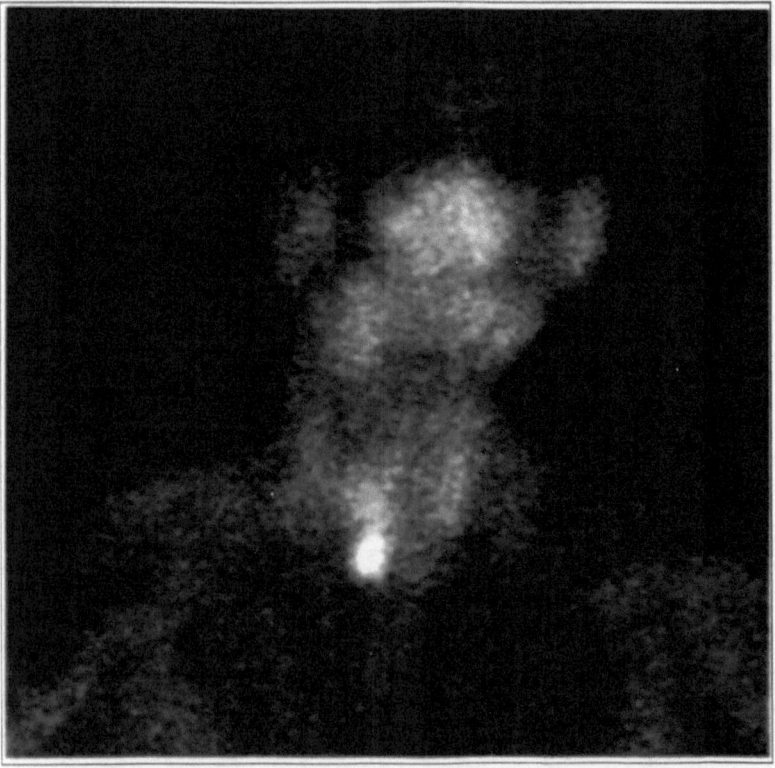

Fig. 8.12. Sestamibi scan in patient with primary hyperparathyroidism. Single focus of intense uptake represents parathyroid adenoma. Scan was obtained one hour following injection and residual thyroid uptake is still visible.

right and left anterior oblique projections. Each image is acquired over a 10 minute interval. The skin overlying the adenoma—normal parathyroid glands are rarely visualized—is marked with an indelible pen. The optimal differential washout occurs at approximately 3 hours postinjection. Our early imaging assures that the surgeon will benefit from the optimal differential washout at the time of actual dissection (Fig. 8.12).

SUMMARY

Lymphoscintigraphy or targeted radioisotope scans are possible using readily available nuclear medicine reagents and equipment. High quality scans performed correctly are essential to a successful radioguided surgical procedure.

REFERENCES

1. Alazraki N. Lymphoscintigraphy and the intraoperative gamma probe. J Nucl Med 1995; 36:1780-1783.
2. Albertini JJ, Lyman GH, Cox C et al. Lymphatic mapping and sentinel node biopsy in the patient with breast cancer. JAMA 1996; 276:1818-1822.
3. Uren RF, Howman-Giles RB, Thompson JF et al. Mammary lymphoscintigraphy in breast cancer. J Nucl Med 1995; 36:1775-1780.
4. Taylor A Jr., Murray D, Herda S, Vansant J, Alazraki N. Dynamic lymphoscintigraphy to identify the sentinel and satellite nodes. Clin Nucl Med 1996; 21:755-758.

8

Pathologic Evaluation of Sentinel Lymph Nodes in Breast Cancer

Ni Ni K. Ku

INTRODUCTION

Lymphatic mapping utilizing technetium labeled sulfur colloid and isosulfan blue dye injections has successfully identified the sentinel lymph nodes (SLNs) and allowed accurate staging.[1,2] This minimally invasive technique may lead to more conservative lymph node dissection for stages I and II breast cancer patients with decreased morbidity and cost savings for the health care system. For most cancers, the nodal status is the single most important prognostic indicator and determines the need in many cases for adjuvant chemotherapy.[3] Micrometastases in the axillary lymph nodes has been reported to be associated with poorer survival.[4] Currently, there is a failure rate of 15% to 20% at five years in node-negative patients which may be attributed to the low detection rate of micrometastases using the routine hematoxylin and eosin (H&E) stain.[5] With the advent of lymphatic mapping, these SLNs may be carefully and thoroughly evaluated for successful identification of micrometastases as SLN involvement determines whether the surgeon will perform a complete lymph node dissection (CLND). Generally, frozen section is not recommended for lymph node tissue as it causes extensive freezing artifacts and may obscure the histology of the tumor or lose the microscopic metastatic tumor cells in the cryostat. In addition, traditional staining methods with H&E alone on sectioned lymph nodes may not detect the micrometastatic tumor cells in a background of millions of lymphocytes. We have developed a protocol utilizing an intraoperative imprint cytology (IIC) of the SLNs to confirm the metastatic tumor in grossly positive and suspicious nodes and to identify micrometastases in grossly negative nodes. IIC has therefore become a crucial part of our intraoperative surgical management, enabling the immediate decision for CLND. IIC is followed by a standard histopathology protocol in concert with ancillary immunohistochemical stain for low molecular weight cytokeratin (CAM 5.2) on grossly negative SLNs.

Radioguided Surgery, edited by Eric D. Whitman and Douglas Reintgen.
© 1999 Landes Bioscience

PATHOLOGY PROTOCOL

The radiolabeled SLNs are submitted to the pathology processing room identified as SLN 1, SLN 2, SLN 3, etc. SLNs are bisected intraoperatively and examined for any gross evidence of metastatic disease. About one-quarter of the SLN is then snap frozen in the operating room for RT-PCR analysis for other protocols. At our institution, all physicians including surgeons, radiologists, pathologists, nuclear medicine staff and intraoperative personnel routinely wear radiation monitoring badges.

The SLNs are dissected free from the surrounding fat and measured. SLNs with a measurement of 5 mm or less in the maximum dimension are bivalved and those greater than 5 mm in diameter are serially sectioned at 2-3 mm intervals to maximize the surface area for IIC. The cut surfaces are carefully examined for gross evidence of metastases and the size of tumor noted. Frozen sections of the SLNs are not done in our institution. IIC utilizes glass slides prelabeled with the patient's initials and SLN number. Imprints are made with a gentle, single touch on each cut surface of the SLN. Imprints of multiple cut surfaces may be prepared on the same glass slide from each SLN without overlapping. The slides are air-dried and stained with Diff-Quik stain (Table 9.1) followed by immediate (intraoperative) interpretation. The turn-around time from specimen receipt until cytologic diagnosis is approximately 5-8 minutes for each SLN. The diagnostic terminologies that are utilized for reporting IIC are: negative, indeterminate and positive. Indeterminate is used when atypical cells of undetermined origin (histiocytic versus epithelial) or rare suspicious cells are identified. From the surgical management point of view, the negative and indeterminate categories are considered negative and only patients with definitively positive SLNs receive immediate CLND.

We have recently developed a technique of intraoperative utilization of immunocytochemical cytokeratin stain on cytologic touch imprints of the SLNs as an adjunct to the Diff-Quik stain. Such technique is extremely useful in supporting the intraoperative diagnosis or detecting micrometastases in lobular or low grade ductal cancers. The VECTASTAIN universal Quick Kit (Vector Laboratories, Burlingame, CA) is used along with a primary monoclonal antibody against low molecular weight cytokeratin (CAM 5.2) (Becton-Dickinson Immunocytochemistry System, San Jose, CA). The turnaround time for such IIC is approximately 16 minutes. Currently, the procedure is performed parallel to the Diff-Quik stain on all SLNs with known lobular and low grade ductal cancers, when the result of Diff-Quik stain is indeterminate and when there is a discrepancy between gross and cytologic findings (grossly positive or suspicious and cytologic imprint negative). Each of the SLN is then placed in separate formalin containers by the nuclear medicine staff and quarantined at room temperature for 48 hours along with the lumpectomy or mastectomy specimen, to allow decay of the technetium labeled probe used for intraoperative lymphatic mapping. The nuclear medicine staff monitors these specimens for background emission and resubmits them to the pathology laboratory when six half-lives of the technetium have dissipated.

Table 9.1. DIFF-QUIK Stain

A. Thoroughly air dry slides.
B. Stain with Diff-Quik solution as follows.
Step 1: Fix for 10-15 seconds in solution I. Drain excess solution onto a paper towel.
Step 2: Dip slides repeatedly for 10-20 seconds in solution II until slides are uniformly coated and turn red-pink. Drain excess stain.
Step 3: Dip slides repeatedly for 10-20 seconds into solution III. Rinse in tap water.
Step 4: Drain excess water. The slides may be examined wet without coverslipping.
Step 5: Mount in resin and coverslip after intraoperative evaluation.

(Editors' Note: Quarantine of pathologic specimens, especially the sentinel nodes, is not universally practiced nor necessarily required, depending on the radionuclide dosage used and the specific radiation safety guidelines of individual institutions. For further details, see the chapter 2). The specimens are then accessioned, the SLNs are submitted entirely and processed for routine histopathological examination with H&E stain. This usually involves one to two blocks per SLN (average one block per SLN). Each block is sectioned at one to three levels per slide for H&E stain. In addition, any SLNs that are grossly and IIC negative for metastases are immunohistochemically stained with a monoclonal antibody against low-molecular weight cytokeratin (CAM5.2), clone NCL-5D3, using the avidin-biotin complex technique with diaminobenzidene chromogen.

PATHOLOGIC FINDINGS

On cytologic touch imprints stained by Diff-Quik technique, the metastatic tumor cells are usually arranged in small to large cohesive clusters, syncytial fragments, as well as single cells (Fig. 9.1). Difficulty in interpretation may occur based upon the low volume of tumor cells regardless of their arrangement or the small size and bland appearance of the metastatic cells, particularly in metastatic lobular or low grade ductal carcinomas. Aggregates of reactive histiocytes may look like epithelial cell clusters on touch imprints, especially following excisional biopsies.

In the permanent histologic sections stained by H&E and cytokeratin immunostain (CKI), the clusters of micrometastatic tumor cells can be seen in the: 1) subcapsular sinuses (most common pattern; 2) medullary sinuses (less common pattern); and 3) interfollicular areas. Diffuse single cell patterns of micrometastases are seen commonly in metastatic lobular carcinoma where H&E is often negative even after multiple levels are examined with scanning magnification, but are found with the immunohistochemical stain for cytokeratin (Fig. 9.2). Micrometastases are arbitrarily defined as tumors of less than 2 mm in size.

Serial sectioning of the lymph node and cutting levels as suggested above from the paraffin blocks to include the nodal capsule are essential since the most common location of micrometastases is in the subcapsular sinuses (Figs. 9.3 and 9.4).

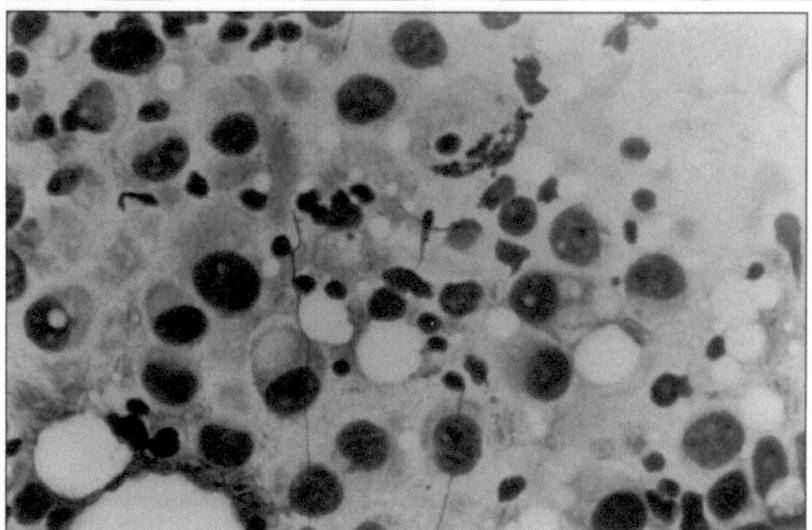

Fig. 9.1. (See Color Insert for color representation). Intraoperative imprint cytology of sentinel lymph node with metastatic ductal Carcinoma as loose clusters and single cells, Diff-Quik stain, X400

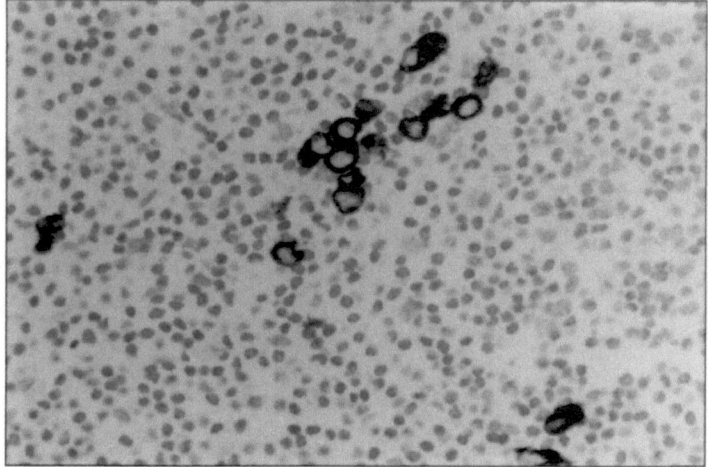

Fig. 9.2. (See Color Insert for color representation). Small microscopic tumor cells in single cell pattern distribution demonstrated only by cytokeratin immunostaining (Hematoxylin & Eosin stain negative), X400

The results of CKI must be interpreted in the context of available data such as tumor/nuclear grade (low grade or high grade), tumor type (ductal or lobular), tumor cell arrangement (cohesive clusters or single cells) and tumor cell distribution. The CKI of SLNs are defined as positive only when the charactistic cytoplasmic staining pattern in the form of "donuts' or "rings" is present, despite the quantity of cells and its arrangement in the given node. Cytokeratin positive cells in the SLN other than metastatic carcinoma are interdigitating dendritic reticulum cells,

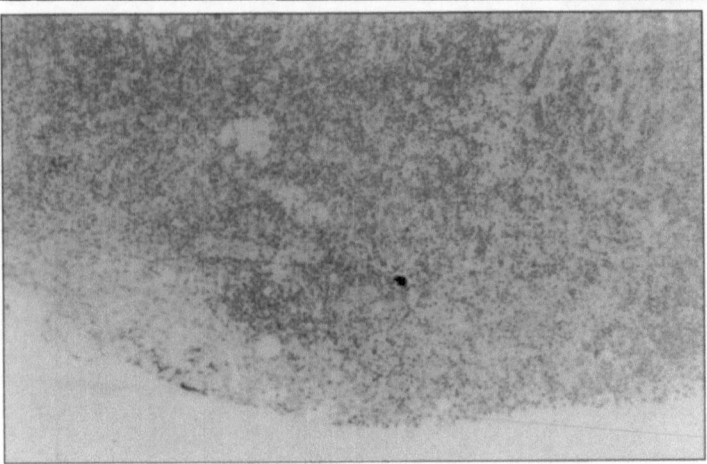

Fig. 9.3. (See Color Insert for color representation). Incomplete sectioning of the lymph nodal capsule missing the subcapsular micrometastases, hematoxylin & e osin stain, X100

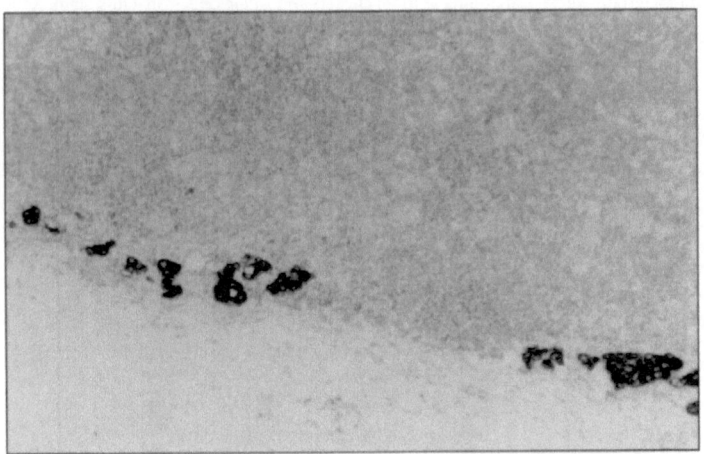

Fig. 9.4. (See Color Insert for color representation). Subcapsular micrometastases high-lighted only cytokeratin immunostaining, X100 (deeper sections of Fig. 9.3)

and benign mammary and adnexal (sweat, sebaceous or apocrine glands) inclusions. Dendritic reticulum cells are distributed around germinal centers in a mesh-like network, and are single cells with small nuclei and dendritic fibrillary cellular processes. Epithelial inclusions and benign nevus cells are very rare and usually located in the nodal capsule or perinodal fat. With experience, these pitfalls can be avoided by comparison of the positively stained areas with the corresponding H&E stained slide and confirming the cells to have a benign histology. We have seen extremely rare cases of a nonspecific staining pattern or aberrant antigen expression appearing as diffuse nuclear and cytoplasmic staining pattern but with mod-

Table 9.2. Errors and potential pitfalls in interpretation of sentinel lymph nodes

	False negative	False positive
Imprint Cytology	sampling error	aggregates of reactive/ atypical histiocytes
	interpretative error	
	low grade ductal or lobular carcinoma	follicle fragments
	low tumor volume	
	single cell distribution	
Permanent Histolgy	incomplete section of nodal capsule	nevus cell rest
	sampling/interpretative error	sinus histiocytosis
	lobular carcinoma	
	low tumor volume	
	single cell distribution	
Tissue CKI	interpretative error	dendritic reticulum cells mammary inclusions adenexal inclusions hemosiderin-laden macrophages

erately weak intensity. This staining is often in an unusual location such as germinal centers of the nodes and usually involves only one or two single cells. Errors and potential pitfalls in interpretation of SLNs in breast cancer patients are given in Table 9.2.

RESULTS

During the period of January 1996 to August 1997, 225 patients with the diagnosis of breast cancer by fine needle aspiration biopsy, core needle biopsy, ABBI or excisional biopsy underwent lymphatic mapping with selective sentinel lymphadenectomy at our institution. Of 397 SLNs (from 225 patients), 16 (from 15 patients) grossly positive or suspicious SLNs were confirmed to be positive (sensitivity 100%) by IIC and those patients were converted to CLND. The results of IIC (by Diff-Quik stain), permanent histology (PH) by standard H&E stain and CKI on the remaining SLNs were compared in Table 9.3, and the significance of CKI upon detecting micrometastases was determined. Of the remaining 381 SLNs (from 210 patients), 22.4% (47 patients) were positive for metastatic carcinoma by H&E and/or CKI, and 77.6% (163 patients) were negative by H&E and CKI. CKI detected micrometastases in 35 and 28 SLNs not identified by IIC (Diff-Quik stain) and H&E, respectively. Seventeen of 47 patients (36.2%) with metastatic disease were detected by CKI only and therefore metastases in such patients would have been missed by H & E alone. Out of 180 patients with H&E negative SLNs, 163 (91.6%) had SLNs that were both H&E and CKI negative while micrometastases was detected in 17 (9.4%) patients by CKI only, enabling to upstage them from N0 to N1 (Table 9.4).

Table 9.3. Correlation between IIC (Diff-Quik), permanent histology and CKI in 381 SNLs

	IIC (Diff-Quik)*	Histology (H&E)	CKI (tissue)
True positive	15	50	68
True negative	254	313	313
False negative	35	28	–
False positive	1	0	–

*IIC not done in 76 cases

Table 9.4. Statistical data including upstaged nodal status by CKI in 210 patients

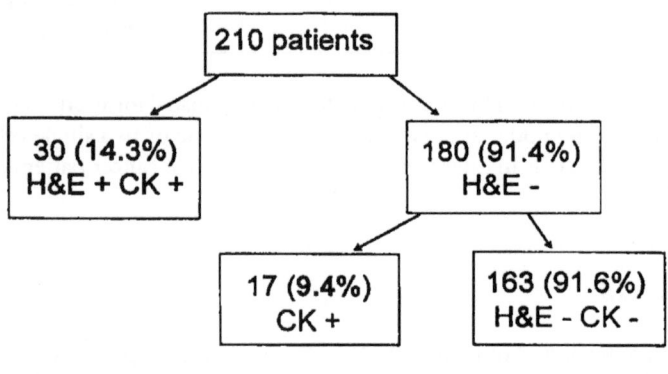

381 Sentinel lymph nodes

210 patients

30 (14.3%)
H&E + CK +

180 (91.4%)
H&E -

17 (9.4%)
CK +

163 (91.6%)
H&E - CK -

August 1997

The above data was recently updated to April 1998 and showed that out of 478 patients with the retrieval of 972 SLNs, 10.6% of patients were upstaged by CKI. No false negative or positive cases were seen with CKI in our series. With a recent introduction of intraoperative CKI on touch imprint of the SLNs in conjunction with the standard Diff-Quik cytologic stain, the detection of micrometastases has become more sensitive and were able to upstage in 27% of patients intraoperatively enabling immediate conversion to CLND.

One false positive cytology in our initial series was associated with reactive/ atypical histiocytes and false negative cases by intraoperative Diff-Quik stain were secondary to sampling error, low tumor cell volume or single cell distribution of lobular or low grade ductal carcinoma cells. Pathological analysis of the nonsentinel lymph nodes in the CLND by H&E stain reveals additional metastatic disease in 45% of patients with both H&E and CKI positive SLNs and only 5% of patients with CKI only positive SLNs. In another recent study, Turner demonstrated that if the SLN is negative by both H&E and CKI, the probability of nonsentinel lymph node involvement is less than 1.0%.[6]

CONCLUSION

Lymphatic mapping with selective sentinel lymphadenectomy allows a detailed pathologic examination of the nodes most likely to contain metastastic disease for accurate staging and detection of micrometastases. Intraoperative identification of micrometastases is critical because it enables the immediate conversion of these patients' surgical treatment to CLND, rather than the CLND being done during a second, deferred, surgical procedure. IIC of the lymph nodes has been used routinely in the diagnosis of hematologic malignancies and also in breast cancer as a useful method in many series.[7,8] Giuliano and Veronesi have reported successful detection of metastases in SLNs by frozen section,[9,10] but we do not recommend these techniques because of potential loss of material in the cryostat, sampling error from low tumor volume, and freezing artifact. In our experience, IIC by Diff-Quik stain is an excellent tool to confirm the SLNs with grossly positive or suspicious (100% sensitive) metastatic disease. Concurrent utilization of intraoperative Diff-Quik stain and CKI appears to be more sensitive in detecting micrometastases intraoperatively based on our preliminary, as yet unpublished, observations. This group of high-risk, traditionally node-negative patients can be more accurately staged with lymphatic mapping and more detailed examination of the SLN. The CKI that we use on the paraffin embedded sections is a low molecular weight (CAM 5.2), clone NCL-5D3, and consists of cytokeratin proteins 8 and 18. This combination of antigens is a very sensitive indicator and is present on all epithelial cells including ductal and lobular epithelial cells of the breast. IIC technique utilizing either Diff-Quik stain or CKI requires intensive training and experience to avoid potential pitfalls in interpretation.

In conclusion, IIC of SLNs stained by Diff-Quik preparation is a straightforward, rapid, and accurate approach to immediate evaluation of lymph nodes at high risk for breast cancer metastases, particularly if those SLN are grossly positive or suspicious. IIC stained by CKI is an extremely useful ancillary technique as an adjunct to Diff-Quik stain in detecting micrometastases particularly in metastatic low grade ductal or lobular carcinoma with low tumor cell volume. Appropriate utilization of Diff-Quik and CKI allows immediate intraoperative staging and cost effective management of breast cancer. SLN mapping technology at our institution, in conjunction with serial sectioning of the SLN and CKI, detected micrometastases in 10.6% of patients not detected by IIC (Diff-Quik) and PH, providing a more accurate nodal staging so that CLND and adjuvant therapy can be given in a selective fashion. These results have encouraged investigators to pursue even more sensitive techniques to uncover micrometastases such as molecular biology technique like reverse transcriptase-polymerase chain reaction (RT-PCR).

Experienced cytopathologists and an active cytopathology service are required to avoid the potential pitfalls of performing and interpreting IIC.

Overall, pathologic examination is a key component of radioguided surgery and lymphatic mapping. One of the proposed advantages of SLN biopsy for breast cancer is that women with negative SLN need not receive complete CLND. However, this surgical decision can only be made with adequate pathology support, both intraoperatively with IIC and postoperatively (serial permanent sections). Performing sentinel node biopsy using only standard, bivalved H&E pathology techniques is an inappropriate application of new technology and is potentially damaging to the patient.

ACKNOWLEDGMENT

I would like to express my sincere appreciation to Santo V. Nicosia, M.D., Chairman of the Department of Pathology at the University of South Florida, Nazeel Ahmad, M.D., Prudence Smith, M.D. and cytopathology residents for their continuous support and participation with daily intraoperative activities, Solange Pendas, M.D. for statistical data, and surgical pathology staff at the Moffitt Cancer Center for their contributions.

REFERENCES

1. Guiliano AE, Kirgan DM, Guenther JM, Morton DL. Lymphatic mapping and sentinel lymphadenectomy for breast cancer. Ann Surg 1994; 220:391-401.
2. Albertini JJ, Lyman GH, Ku NNK, Cox CE, Nicosia SV, Reintgen DS et al. Lymphatic mapping and sentinel node biopsy in patients with breast cancer. JAMA 1996; 276:1818-1822.
3. Statman R, Giuliano AE. The role of the sentinel lymph node in the management of patients with breast cancer. Advances in Surgery, Vol. 30, Mosby Year Book, Inc. 1997.
4. Anderson BO, Austin-Seymore M, Gralow J, Moe RR, Byrd DR. A multidisciplinary approach to locoregional management of the axilla for primary operable breast cancer. Cancer Control 1997.
5. Dowlatshahi K, Fan M, Snider HC. Lymph node micrometastases from breast carcinoma. Reviewing the dilemma. Cancer 1997; 80(7):1188-1197.
6. Turner RR, Ollila DW, Krasne DL, Guiliano AE. Histopathological validation of the Sentinel Lymph Node hypothesis for breast carcinoma. Ann of Surg 1997; 226:271-276.
7. Fischer CJ, Boyle S, Burke M, Price AB. Intraoperative assessment of nodal status in the selection of patients with breast cancer for axillary clearance. Br J Surg 1993; 80:457-458.
8. Hadjiminas J, Burke M. Intraoperative assessment of nodal status in the selection of patients with breast cancer for axillary clearance. Br J Surg 1994; 81:1615-1616.
9. Guiliano AE, Dale PS, Turner RR et al. Improved axillary staging of breast cancer with sentinel lymphadenectomy. Ann Surg 1995; 222(3):394-401.
10. Veronesi U, Paganelli G, Galimberti V et al. Sentinel node biopsy to avoid axillary dissection in breast cancer with clinically negative lymph nodes. Lancet 1997; 349:1864-1867.

Pathologic Evaluation of the Sentinel Lymph Node in Malignant Melanoma

Jane L. Messina

INTRODUCTION

Successful lymphatic mapping and sentinel lymph node (SLN) biopsy requires a specialized but multidisciplinary approach, utilizing the surgeon, nuclear medicine physician, and pathologist. With sentinel lymphadenectomy rapidly becoming the standard of care for patients with melanoma at risk for metastases, pathologists are encountering these specimens with increasing frequency in their daily practice. The pathologic status of the sentinel lymph node is pivotal to the patient's care because it provides staging information that dictates the need for further therapy. Therefore, defining a method for detailed assessment of this tissue is of the utmost importance. Our standard pathology protocol for SLN includes complete submission of all tissue with routine use of immunohistochemical staining for S-100 protein, and intraoperative touch preparation cytology as an adjunct technique in SLN grossly suspicious for metastatic disease. This chapter reviews the literature concerning pathology of sentinel nodes in malignant melanoma patients and describes our results in over 400 melanoma patients to date.

REVIEW OF THE LITERATURE

In the presentinel node era, lymph nodes from melanoma patients were excised as part of complete nodal basin dissections, either for therapeutic purposes or as an elective procedure. Pathologic evaluation procedures often entailed only submission of a representative section from each node identified. A limited amount of prognostic information could be gathered from these dissections. Lymph node parameters demonstrated to have an adverse effect on disease-free and overall

Radioguided Surgery, edited by Eric D. Whitman and Douglas Reintgen.
© 1999 Landes Bioscience

survival include: the number of positive nodes (> 5 vs 2-4 vs 1), clinical vs micro-scopic nodal involvement (80% vs 36-50% five year survival), and extracapsular extension.[1-3] However, even in the mid-1980s it was recognized that the incidence of nodal involvement in elective dissections may be underestimated with H&E stains alone. Cochran et al demonstrated a 14% increase in the incidence of metastatic melanoma detected in elective node dissections using S-100 immunostaining.[4]

With the advent of selective lymphadenectomy, pathologists have recognized that a detailed method for assessing these nodes is needed, but there is little con-currence in the general literature as to exactly what constitutes an ideal protocol. The standard method of evaluating lymph nodes for metastatic tumor involves submitting 1-2 representative sections from the center of the lymph node for H&E staining. This method is sensitive enough to detect 1 abnormal cell in a back-ground of 10^4 normal cells.[5] This sensitivity can be increased 10-fold by using immunohistochemical staining.[6] Immunohistochemical stains which have been particularly useful in melanoma diagnosis include S-100 and HMB-45. S-100 is an acidic, calcium-binding protein so named because of its solubility in 100% ammonium sulfate. It was first identified in mammalian brain tissue, and subse-quently identified in tumors of neural origin. Extensive studies have now shown this protein to be present in a wide variety of non-neurogenic tissues, as well are their corresponding tumors.[7] S-100 antibody is 95-100% sensitive in detecting both primary and metastatic melanoma.[8] HMB-45, a marker of melanosomes, is 92% sensitive in diagnosing primary melanoma[9] but in our experience it is ap-proximately 50% sensitive in diagnosing micrometastatic malignant melanoma (unpublished observations).

The usefulness of immunohistochemistry in studying sentinel lymph nodes was demonstrated in an early study of 15 sentinel node negative patients with nodal basin recurrence. Retrospective evaluation of these patient's sentinel nodes using serial sectioning, S-100, and HMB-45 immunostains confirmed the pres-ence of micrometastases in 66% of these "negative" sentinel nodes.[10] A more re-cent study confirmed these findings.[11] In this abstract, the immunostains were instrumental in identifying the occult metastases, while serial sections were useful in only one case.

PATHOLOGIC PROTOCOL FOR HANDLING SENTINEL NODES

With these issues in mind, we instituted the following protocol in October 1995 (Table 10.1). Following harvesting from the patient, each SLN is placed in a separate formalin container and held at room temperature for at least 48 hours (six 1/2 lives) to allow decay of the technetium-labeled sulfur colloid used for intraoperative mapping. Specimens are then accessioned and processed in the pathology laboratory in the same manner as routine, nonradioactive specimens. Gross examination is performed, and the presence or absence of vital blue dye,

Table 10.1. Algorithm for pathologic evaluation of sentinel node

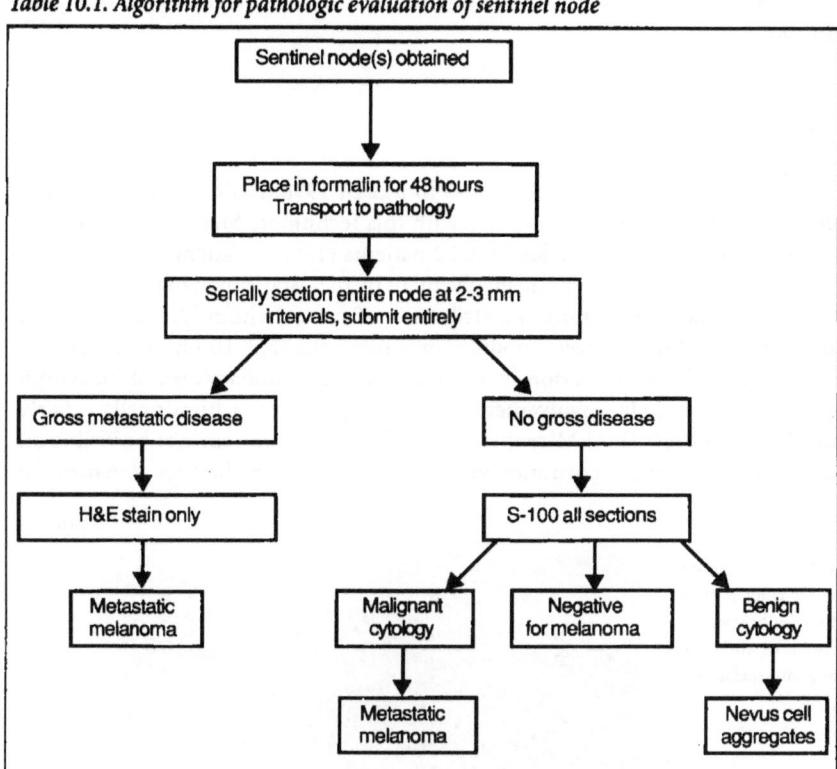

melanin-like pigment, and tumor nodules are noted. The lymph node is sectioned at 2-3 mm intervals and entirely submitted for paraffin embedding. This usually entails submission of one to eight cassettes of tissue per SLN (average of 1.5 blocks/node). One glass slide per block is cut and stained with H&E; this slide may contain 1-3 lymph node profiles, depending on the size of tissue in the block. Additionally, all blocks of lymph nodes that are grossly negative for melanoma metastases are stained with S-100 protein antibody using the avidin-biotin complex immunoperoxidase technique with diaminobenzidine chromogen. When completion dissections are performed (after a positive SLN biopsy), the lymph nodes are entirely embedded and sections stained with H&E. We have studied a group of 20 complete node dissections with S-100 immunostaining as well (see results).

Occasionally, sentinel nodes may contain small areas of brown-black pigmentation without frank tumor nodularity, and as such be suspicious for metastatic melanoma. In this selected group of patients, we may perform touch preparations for cytologic examination before fixation at the time of lymphadenectomy. The

SLN is bivalved and each half is touched with a slide, which is stained with Diff-Quik.

RESULTS

During the period of October 1995 to March 1998, 1005 SLN from 405 patients at our institution were studied with this technique. Metastatic melanoma was detected in 92 lymph nodes from 72 patients (18% of patients).

Four patterns of lymph node involvement by metastatic malignant melanoma are seen: 1) The most common distribution of tumor within lymph nodes is a **subcapsular nodule** or several nodules of tumor cells (Fig. 10.1); 2) Tumor may less commonly be seen as a dominant nodule in the **medullary** area of the lymph node; 3) Diffuse **nodal involvement** by small aggregates of tumor cells is fairly uncommon (Fig 10.2); 4) **Micrometastatic disease**, which may not be visible on initial H&E sections, is encountered in two situations. In the first scenario, the

Fig. 10.1. (See Color Insert for color representation). Subcapsular aggregate of metastatic malignant melanoma.

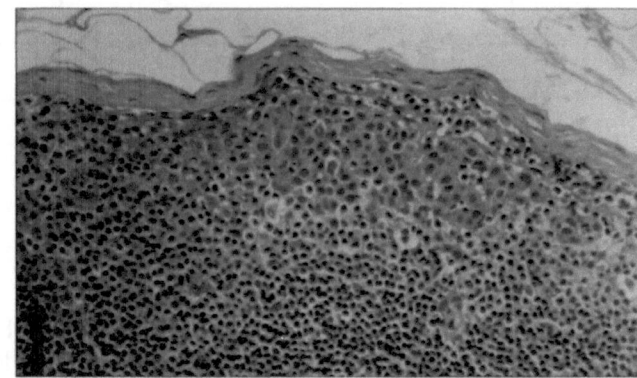

Fig. 10.2. (See Color Insert for color representation). Diffuse distribution of metastatic malignant melanoma.

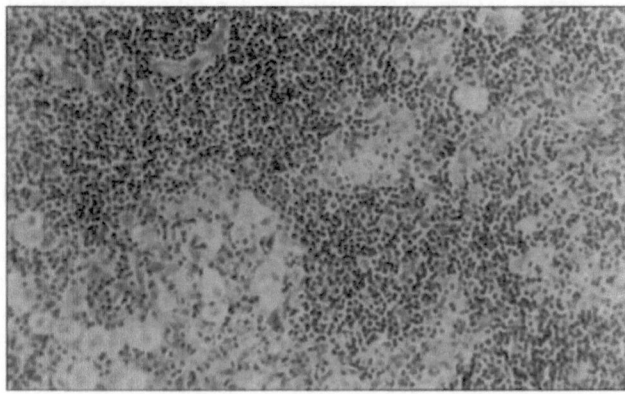

tumor volume is so small that it is not present in initial sections but only detected after deeper sections through the block are performed for immunostaining. In the second situation, the tumor is only identified on immunostaining, but on retrospective review is present in initial sections in amounts too low to be appreciated on scanning magnification (Figs. 10.3, 10.4). The techniques described above are especially useful in evaluating micrometastatic disease.

Cytologic features of the typical melanoma tumor cell are well known: the cells are large, epithelioid, and display ample eosinophilic to brown-black cytoplasm, a prominent nucleolus, and occasional nuclear pseudoinclusions.

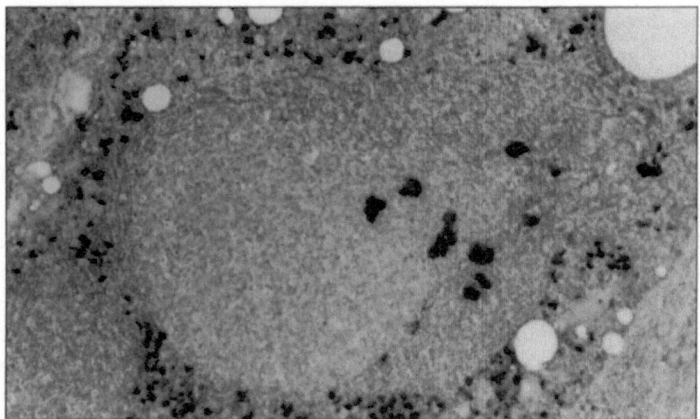

Fig. 10.3. (See Color Insert for color representation). Rare clumps of strongly S-100 positive cells of metastatic malignant melanoma, scattered weakly positive single dendritic cells.

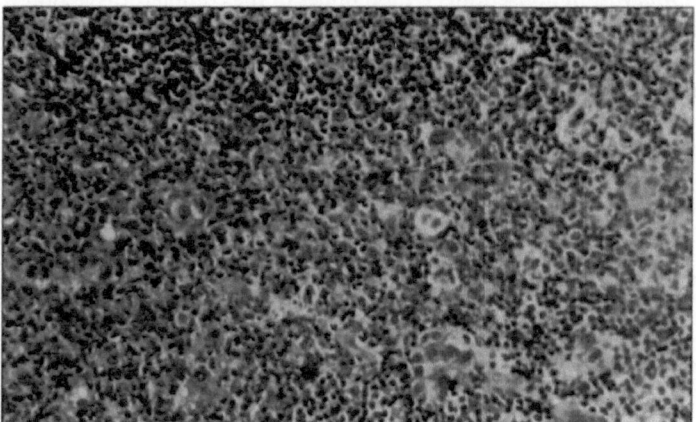

Fig. 10.4. (See Color Insert for color representation). Histologic section corresponding to Fig. 3, showing rare aggregates of pigmented, cytologically malignant cells.

Multinucleated tumor cells and bizarre mitotic figures are frequent. Cytoplasmic melanin formation is not often readily visible on H&E sections. The diagnosis of micrometastatic malignant melanoma rests on finding clumps or aggregates of strongly S-100 positive cells within the lymph node, which on retrospective review of the original H&E section or deeper H&E sections are found to demonstrate these malignant cytologic features.

In our study population metastatic melanoma was detected in 95 lymph nodes from 72 patients. In 62 of these lymph nodes (from 49 patients), metastatic disease was detected on the H&E stain alone. S-100 immunostaining revealed an additional 31 positive nodes from 23 patients, increasing the yield of detected disease by 33% (32% of all node positive patients). In three lymph nodes from three of these latter patients (4%), the tumor was not present on the original H&E slide and was only seen on the deeper section cut for immunostaining (Table 10.2). Thus, in 23 patients (32% of patients with positive sentinel lymph nodes), the diagnosis of metastatic melanoma would have been missed on H&E sections alone and was only picked up by S-100 immunostaining. Overall, 19/405 patients (4.7% of total study population) were upstaged using these special techniques. These 19 patients comprised 7% of the node-negative population.

Touch preparations were performed on 23 lymph nodes from 13 patients and results compared to the final pathology using the above techniques. In three lymph nodes from three patients, the cytology was interpreted as negative or suspicious but not diagnostic for metastasis, while the corresponding histology revealed metastatic melanoma. In the remaining 19 lymph nodes (10 patients), there was agreement between the cytology and histology (sensitivity of 62%, specificity of 100%).

In all patients with a positive sentinel lymph node, a completion nodal dissection of the involved basin was offered, and it was performed in 61 patients (85% of SLN positive patients). Pathologic analysis of these lymph nodes revealed additional disease in 6 (10%) of these patients. Overall, the incidence of metastatic melanoma within nonsentinel lymph nodes of the complete dissection was 1.4%. This is in contrast to the frequency of metastatic disease in the SLN, which is 18%. No additional disease in the completion node dissection specimens was uncovered using S-100 immunostaining in the 20 cases subjected to this technique.

Table 10.2. Histologic findings in sentinel lymph nodes

	# Patients	% Patients	# Lymph Nodes	% Lymph Nodes
H&E first	47	65%	62	65%
H&E only	2	3%	2	2%
S-100 first	23	32%	31	33%
Total	72	100%	95	100%

DISCUSSION

Identification of metastatic disease in SLNs plays a critical role in the management of melanoma patients, and the techniques used must be both sensitive and specific. The specificity of S-100 is not high; S-100 antigen is present on a wide variety of normal and abnormal lymph node constituents (Table 10.3). The most numerous of these are the interdigitating reticulum cells (IRC) of the interfollicular regions. These cells form a mesh-like network surrounding germinal centers. They are seen as single cells with small nuclei, the same size as lymphocyte nuclei, and dendritic cellular processes (Fig. 10.5). IRC stain lightly with S-100. In contrast, metastatic melanoma forms cohesive cellular aggregates of strongly S-100-positive tumor cells, which are much larger than the surrounding lymphocytes. These cells

Table 10.3. Distribution of S-100-positive cells in lymph nodes

S-100 Positive Cells	Distribution
Interdigitating Reticulum Cells	Around Germinal Centers, Fine Net-like Distribution
Adipose Tissue	Perinodal Fat
Peripheral Nerves	Intranodal and Perinodal Connective Tissue
Benign Nevus Cell Aggregates	Small, Nested Groups within Lymph Node Capsule or Fibrous Trabeculae
Metastatic Malignant Melanoma	See Text

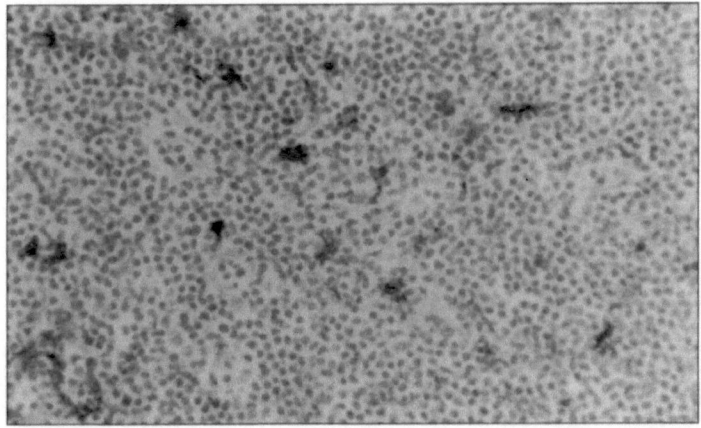

Fig. 10.5. (See Color Insert for color representation). Typical pattern of interdigitating reticulum cells.

must be positively identified as having malignant cytologic features on the corresponding H&E stain before they are diagnosed as melanoma.

Another S-100 positive cell found in lymph nodes is the benign nevus cell aggregate (NCA). Small groups of S-100 positive, cytologically benign melanocytes are found in the lymph node fibrous capsule or trabeculae in < 1-22% of lymphadenectomy specimens.[12] The reported frequency of benign NCA in the literature ranges from 33% to 7.3%.[13,14] Numerous theories have been expounded concerning the origins of these cells, from errant migration from the neural crest to dislodgment and "benign metastasis" from cutaneous nevi. These cells can be distinguished from malignant melanoma by their completely benign cytology (Figs. 10.6, 10.7). NCA were found in 6.3% of sentinel lymph nodes in our study population.

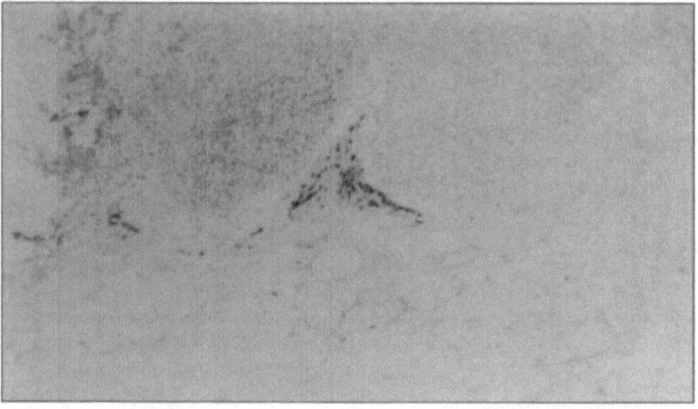

Fig. 10.6. (See Color Insert for color representation). Intracapsular aggregate of S-100 positive cells of a benign nevus cell aggregate.

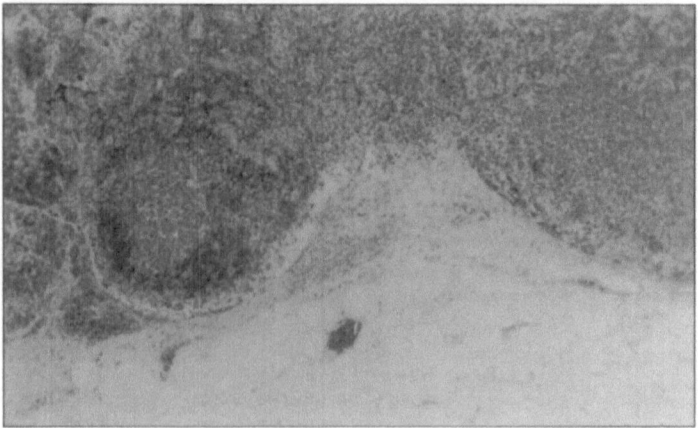

Fig. 10.7. (See Color Insert for color representation). Histologic section corresponding to Fig. 10.6, showing benign cytology of nevus cell aggregate.

In contrast to S-100, the HMB-45 antibody, while quite specific (95%), has been less sensitive (45-65%) in detecting micrometastatic disease in our patients. HMB-45 immunostaining is useful in cases where S-100 positive cells are found within the lymph node, but demonstrate equivocal cytologic atypia. If questionable areas are found to be HMB-45 positive, they are considered to represent metastatic melanoma.

Since histologic techniques are not 100% sensitive in detecting all micrometastatic disease, several other methods have been investigated to this end. A cell culture technique using portions of lymph nodes from melanoma nodal dissections, followed by immunohistochemical staining, increased detection of occult nodal metastases by 31%.[15] The reverse-transcriptase polymerase chain reaction technique for detection of tyrosinase revealed that 47% of histologically negative lymph nodes contained the RNA signal for tyrosinase.[5] This has recently been shown to be of prognostic significance; the recurrence rate for these patients is 13%, compared to 2% for lymph nodes negative both histologically and by PCR analysis.[16] Serial sectioning has also been advocated, increasing detection of metastatic melanoma by up to 24%.[17]

Our results are in concurrence with these data. Immunohistochemical staining of SLNs significantly increases the detection of melanoma metastases. The antibody of choice for this procedure is S-100, because with a sensitivity of 95-100%, it is a perfect screening tool. Pathologic interpretation of S-100 stained lymph nodes may be difficult because the antibody lacks specificity, staining a number of normal lymph node counterparts. With experience, however, such pitfalls can be avoided by comparison of the positively-stained areas with the corresponding H&E-stained slide and confirming the benign cytology of the cells in question.

Other pitfalls in the diagnosis of metastatic melanoma include melanin-laden macrophages and foamy macrophages. Melanin-laden macrophages (melanophages) are frequently found in lymph nodes draining the sites of chronic cutaneous inflammation (dermatopathic lymphadenopathy). Their existence, therefore, in lymph nodes draining the site of a malignant melanoma is not surprising. The presence of large nodal aggregates of melanophages is a cause for interpretative caution. S-100 is extremely useful in detecting the presence of viable melanoma cells in such lesions, as the antibody does not stain most melanophages. In heavily pigmented lymph nodes, however, a red immunoperoxidase chromogen (aminoethylcarbazine) is used in lieu of the standard brown end product. Expansion of the subcapsular sinuses by histiocytic infiltrates may be especially prominent in lymph nodes draining regions of joint prostheses or silicone breast implants. This may occasionally be so exuberant as to resemble metastases of balloon cell melanoma. However, careful high-power examination will reveal that these histiocytes display benign nuclear features.

In summary, the technique of lymphatic mapping and SLN biopsy is an effective method of accurately staging melanoma patients and identifying those who may benefit from further surgery and/or adjuvant chemotherapy. The typical SLN pathology specimen consists of 1-2 lymph nodes, which are the most likely to contain metastatic tumor and a detailed pathologic examination using immunohistochemistry is both feasible and practical. We have demonstrated that thorough evaluation of sentinel lymph nodes requires full utilization of the tissue submitted to the pathologist. This evaluation includes complete embedding of the entire lymph node as well as routine immunohistochemical staining for S-100 in all nodes grossly negative for tumor. These techniques will increase the yield of detected disease by 39%. Although our study sample was small, the technique of touch preparation of these lymph nodes appears to be useful in nodes, which are grossly suspicious for metastases. Of our study population of 405 patients, only six patients with positive sentinel lymph nodes (1.4%) had additional metastatic disease in the completed resected lymphatic basin. All of these patients had a primary melanoma > 3.00 mm in thickness. Because of this low incidence of additional disease in the basin outside of the sentinel lymph node, at this time S-100 immunostaining of the entire dissection specimen does not appear to be warranted. Further studies will be needed to determine the prognostic significance and clinical correlation of such occult, micrometastatic disease in patients with malignant melanoma.

REFERENCES

1. Buzaid AC, Tinoco LA, Jendiroba D et al. Prognostic value of size of lymph node metastases in patients with cutaneous melanoma. J Clin Oncol 1995; 13:2361-8.
2. Coit DG, Rogatko A, Brennan MF. Prognostic factors in patients with melanoma metastatic to axillary or inguinal lymph nodes. A multivariate analysis. Ann Surg 1991; 214(5):627-636.
3. Callery C, Cochran AJ, Roe DJ et al. Factors prognostic for survival in patients with malignant melanoma spread to the regional lymph nodes. Ann Surg 1982; 196:69.
4. Cochran AJ, Wen DR, and Morton DL. Occult tumor cells in the lymph nodes of patients with pathological stage I malignant melanoma. Am J Surg Pathol 1988; 12(8):612-618.
5. Wang X, Heller R, Cruse CW, VanVoorhis N, Glass LF, Fenske NA, Berman C, Leo-Messina J, Rapapport D, Wells K, DeConti R, Moscinski L, Stankard C, Puleo C, Reintgen DS. Detection of submicroscopic lymph node metastases with polymerase chain reaction in patients with malignant melanoma. Annals Surg 1994; 220(6):768-774.
6. Morton DL, WEN DR, Wong J et al. Technical details of intraoperative lymphatic mapping for early stage melanoma. Arch Surg 1992; 127:392-399.
7. Kahn HJ, Marks A, Thom H, Baumal R. Role of antibody to S100 protein in diagnostic pathology. Am J Clin Pathol 1983; 79:341.
8. Gaynor RB, Herschman HR, Irie R et al. S-100 protein: A marker for malignant melanoma. J Clin Pathol 1985; 38:7-15.
9. Fernando SS, Johnson S, Bate J. Immunohistochemical analysis of cutaneous malignant melanoma: comparison of S-100 protein, HMB-45 monoclonal antibody and NKI/C3 monoclonal antibody. Pathology 1994; 26(1):16-19.
10. Gershenwold J, Thompson W, Mansfield P et al. Patterns of failure in melanoma patients after successful mapping and negative sentinel node biopsy. Ann Surg, in press.

11. Colome-Grimmer MI, Gershenwald J, Ross MI et al. The negative sentinel lymph node in Stage I-II melanoma patients. Retrospective study of 20 cases after recurrence and of 41 cases prospectively. Abstract. J Cutan Pathol 1997; 24(2):91.

12. Carson KF, Wen DR, Li P, Lana A, Bailly C, Morton DL, Cochran AJ. Nodal Nevi and Cutaneous Melanomas Am J Surg Pathol 1996; 20(7):834-840.

13. Bautista NC, Cohan S, Anders KH. Benign melanocytic nevus cells in axillary lymph nodes. A prospective incidence and immunohistochemical study with literature review. Am J Clin Pathol 1994; 102:102.

14. Yazdi HM. Nevus cell aggregates associated with lymph nodes. Immunohistochemical Observations. Arch Pathol Lab Med 1985; 102:1044.

15. Heller R, King B, Baekey P, Cruse W, Reintgen D. Identification of submicroscopic lymph node metastases in patients with malignant melanoma. Semin Surg Oncol 1993; 9:285.

16. Shivers S, Wang X, Cruse W, Fenske N, DeConti R, Messina J, Glass LF, Berman C, Reintgen DS. Molecular Staging of Melanoma, JAMA, submitted.

17. Cochran AJ, Wen DR, Morton DL. Occult tumor cells in lymph nodes of patients with pathological stage I malignant melanoma. Am J Surg Path 1988; 12:612.

10

The Technique of Minimally Invasive Radioguided Parathyroidectomy (MIRP)

James Norman

Primary hyperparathyroidism (HPTH) is the result of a single adenoma in 87-92% of all cases and is cured long-term following the removal of this one gland.[1] Despite this fact, most surgeons continue to perform a complete bilateral neck exploration for patients with primary HPTH. The purpose of this comprehensive dissection is to identify and even biopsy each gland so all hyperfunctional parathyroid tissue is identified and resected while normal glands are left behind. This conviction is a historical reflection of the inability of preoperative testing to accurately distinguish those patients harboring a single diseased gland from the 8 to 13% with multiple adenomas or four gland hyperplasia.

The advent of the sestamibi scan in the early 1990s has changed the management of primary HPTH for many surgeons. A recent meta-analysis of more than 6000 patients has shown preoperative sestamibi scanning to have a sensitivity of approximately 90% and a specificity approaching 100% in identifying patients with a single adenoma.[1] This suggests that the vast majority of all parathyroid adenomas will be identifiable using this technique. More important is the fact that when a single adenoma is identified, the scan is almost always correct.

Using the technique of sestamibi scanning followed immediately by surgical exploration, we have shown that the radioactive gland can be rapidly identified by using a miniature gamma detection device intraoperatively to guide the dissection.[15] This technique allows a smaller more directed approach to the adenoma which is easily completed using local anesthesia in a true outpatient setting. The advantages of this technique are greater than they appear on the surface. The smaller operative field dictates that a smaller incision can be used which has a tremendous influence on patients and referring physicians alike. The use of local anesthesia also has a positive influence on the patient's view of surgery preoperatively, but it also more readily allows for them to be sent home almost immediately after the procedure. Use of the probe in the operating room allows the surgeon to pro-

Radioguided Surgery, edited by Eric D. Whitman and Douglas Reintgen.
© 1999 Landes Bioscience

ceed with confidence toward the adenoma allowing him/her to close the wound and call for the next patient without waiting for frozen section histologic diagnosis or searching for other glands. The cost savings of this approach has been examined by our group previously, noting that it becomes cost-effective to perform sestamibi scans on all patients when more than 50% of patients can undergo a unilateral neck exploration (without radioguidance). When patients can undergo a MIRP, this break-even point decreases to only 35%. Our experience shows that approximately 84% of all patients with sporadic primary HPT can have a MIRP.

PATIENT SELECTION AND ANTICIPATED RESULTS

Once the diagnosis of primary HPTH has been confirmed, patients are counseled regarding the standard versus minimally invasive approach. We do not perform a sestamibi scan or any other localizing studies prior to the day of the operation. Patients are scheduled for an operation and the operative technique used (standard vs MIRP) will be dictated by the results of the sestamibi scan an hour or two prior to surgery. Because of the critical nature of timing (discussed below), we suggest that surgeons learning this technique begin by choosing their patients for MIRP through the use of preoperative scanning a week or two prior to the operation. Then, on the morning of surgery, a second sestamibi scan is performed such that the patient is in the operating room about 2-2.5 hours after the reinjection of the radiopharmaceutical. Once the nuclear medicine department, the surgeon, and the entire OR team realize the critical nature of timing, then the screening sestamibi scans should be eliminated. The goal for surgeons using this technique is to scan patients only once—1-2 hours prior to surgery.

Since not all patients will demonstrate a single adenoma on sestamibi (89% have an adenoma and the typical sestamibi sensitivity in the literature is 90%), they are informed that some will not be candidates for minimal resection. Figure 11.1 details the selection of all patients with primary HPTH for MIRP versus standard bilateral exploration. This diagram is given to each patient when counseling them preoperatively and explained in detail. If a single adenoma is found on the preoperative scan, then a MIRP is performed. If no localization occurs, a standard bilateral exploration will be performed at that time. Because of the dramatically increased referral pattern associated with the use of this minimally invasive approach, an occasional patient will elect not to undergo a standard exploration if a single adenoma fails to visualize (the morbidly obese, extremes of age with confounding medical problems).

OPTIMAL TIMING BETWEEN SESTAMIBI
AND OPERATIVE EXPLORATION

There are two critical determinants of successful MIRP. The first is the quality of the preoperative sestamibi which allows the surgeon to operate with conviction

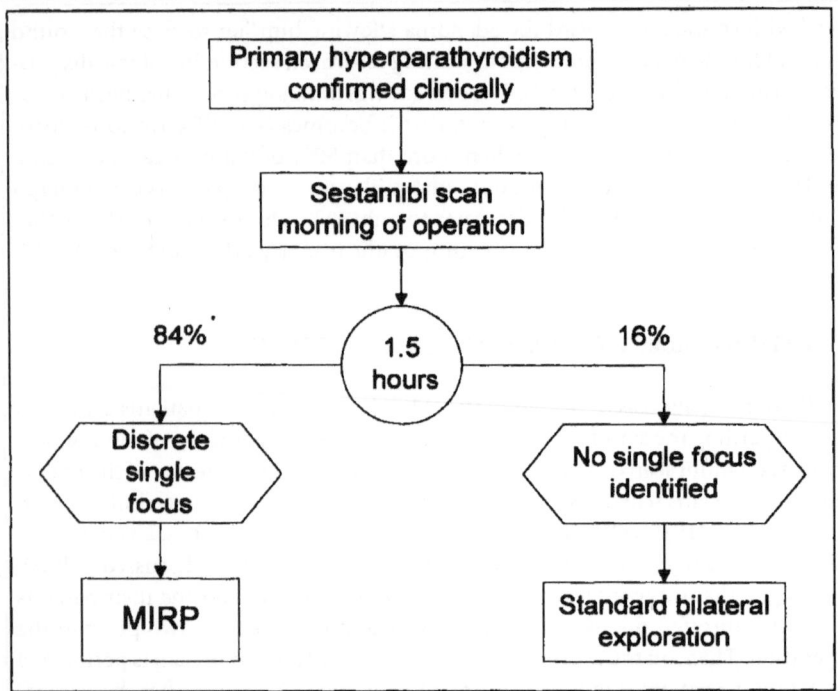

Fig. 11.1. Selection of operative procedure for patients with primary hyperparathyroidism for MIRP. Following clinical conformation of primary HPTH, the patient is scheduled for the operating room with a sestamibi scan to precede the OR by 3 hours. The majority of patients will localize to a single adenoma and are candidates for out-patient minimally invasive resection under local anesthesia. Those who do not localize on scanning are explored at that time using standard techniques. No other localizing studies are performed prior to the morning of surgery.

on a single area of the neck without the need for identification of other "normal" glands while making the chances of persistent disease due to a missed adenoma extremely low. The second critical determinant for success became obvious only after we had accumulated sufficient experience with this technique to realize that there was a "window" of optimal timing between injection of the radiopharmaceutical and using the probe in the operating room.

Several factors dictate the optimal time between the sestamibi scan and intraoperative nuclear mapping/resection. The first consideration is the half-life of the radiopharmaceutical which is approximately 6.0 hours. This dictates that the two procedures must be carried out on the same day. The most important consideration, however, is the relative speeds at which the thyroid and parathyroid "wash out" their nuclear tag. Since the thyroid will lose its initial uptake of sestamibi-Tc^{99} at a faster rate than a hyperactive parathyroid gland, the optimal situation occurs when the thyroid has washed out and the parathyroid remains relatively radioactive. Only when there is differential activity between the thyroid and the

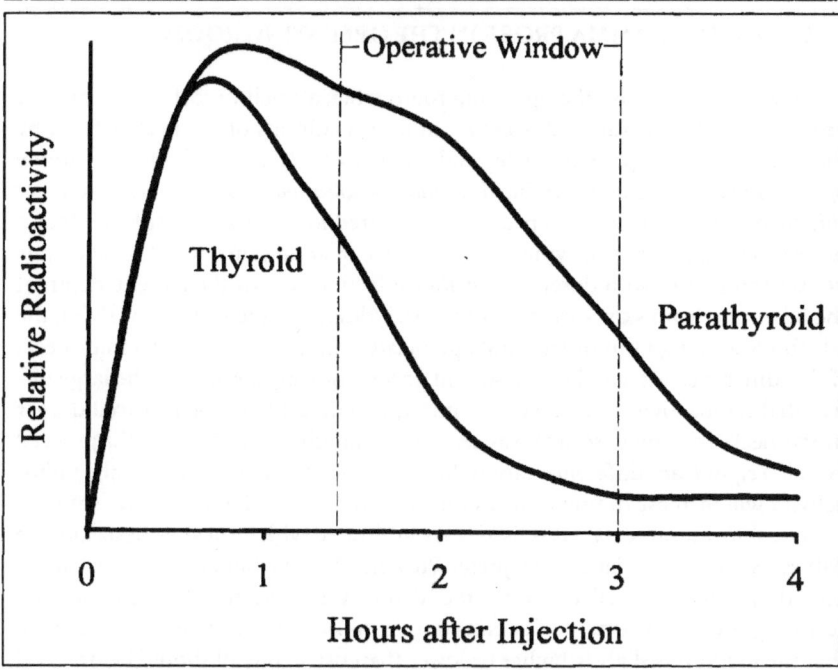

Fig. 11.2. Operative Window for MIRP. The optimal timing for performing radioguided parathyroidectomy is determined by several factors but none is more important than the length of time which has expired since the patient was injected with radiopharmaceutical. The gamma probe will be most accurate at a time when the thyroid has lost more of its radioactivity than the parathyroid. Once beyond 3.5 hours, the radioactivity in the parathyroid has washed out to such a degree that it is difficult for the probe to distinguish it from background.

parathyroid adenoma can the gamma probe be helpful. We have found that this situation occurs within a window between 1.5 and 3.0 hours. Typically, 1.5-2.5 hours is ideal for the vast majority of patients and we discourage an elapsed time over 3.5 hours. We have had the occasion to try this technique several times between 4.0-6.5 hours after technetium injection. In this setting, no differential radioactivity could be found and a true MIRP was not possible. We typically schedule the sestamibi injection for 07:30 with the operating room scheduled for 09:30. If a second case will follow, the injection should follow the first case by about 1.0-1.5 hours. The details of how we perform sestamibi scanning prior to MIRP can be found on-line at **www.EndocrineWeb.com** using the keyword **sestamibi**.

We believe that the risk of recurrent nerve injury is not increased and may even be less for the MIRP procedure. Dissection around the ipsilateral nerve is limited because the adenoma is localized with the gamma probe. The contralateral side is obviously without risk. We have not had this complication in our several hundred case experience, but inform our patients that the expected incidence is comparable to standard exploration at about 1%.

USE OF THE GAMMA PROBE IN THE OPERATING ROOM

When positioned on the operating room table, a small (9 to 14 mm) gamma probe is used to measure radioactivity in four quadrants of the neck defined by the upper and lower poles of the thyroid on each side. *Note that the first generation of probes which were large, unshielded, and noncolumnated are not very useful for this operation.* The new small probes are required for successful implementation of this technique. There is typically a difference of approximately 100-400 counts per second (varies with dosage, time after injection, size of the patient, depth of the adenoma from skin surface, and the size/design of the probe) overlying the adenoma, although this often is not appreciated for a deep lesion in a thick neck. If this difference is seen, the case should go very quickly because of the large differential radioactivity which will become apparent as the probe is placed deeper in the neck. If a single source was seen on sestamibi, cases should still proceed even if a significant difference cannot be detected at the skin level since the radioactivity will increase dramatically as the dissection nears the radioactive source.

The skin and subcutaneous tissues are infiltrated with local anesthesia and the patient is given IV sedation (we prefer Propofol™). The initial incision is placed according to the expected location of the adenoma as determined by both sestamibi scanning and measurement of gamma emissions on the skin. This will necessitate that the incision is slightly higher or lower than usual, but all should be oriented transversely to allow extension as needed, or even conversion to bilateral exploration if necessary. Superficial adenomas (at the level of the thyroid lobe) can be removed through a 2.0-2.5 cm incision. Those adenomas lying in the tracheo-esophageal groove, however, usually require a 3 cm incision.

Sub-platysma flaps are created 1.0-1.5 cm in all directions and held open with a small self-retaining retractor. Radioactivity is again quantitated in all four quadrants. The strap muscles are now separated along the midline and another self-retaining retractor is placed at 90° to the original to hold these muscles apart. The dissection is carried deeper as directed by increasing gamma counts to locate the radioactive gland. Beyond this point, blunt dissection should be used exclusively to prevent damage to small vascular or nervous structures. Cautery below the strap muscles can usually be avoided completely, but when needed, any deep cautery should be of the bipolar type. The recurrent laryngeal nerve is examined if the radioguided dissection includes it in the operative field. We do not make a specific point of locating the nerve during every case, but are constantly aware of its anatomical relationships.

The adenoma is located by continued and frequent use of the probe to direct the dissection. When placing the probe deep in the neck, it must be remembered not to aim it directly at the heart as the sestamibi-Tc^{99} is also used as a cardiac imaging agent and will give false-positive readings. Once identified, the adenoma is teased from its surrounding tissues bluntly and elevated to reveal its single pedicle which is ligated with a hemoclip and transected. A drain is not needed. At no time should safety be compromised by a hesitancy to extend the incision or even conversion to general anesthesia if necessary. When using the probe to its full potential there will be very little unnecessary dissection. Our experience has shown that as many as 35% of all parathyroid adenomas can be removed without even exposing the thyroid gland.

We routinely send MIRP patients home within an hour or two of the procedure. Those patients with significant underlying medical problems are kept overnight, but this has been necessary only 9 times in over 200 cases. Advanced age alone is not reason enough to preclude an outpatient approach if all has gone well.

There are several important points following removal of the adenoma which combine to acknowledge it as the anticipated source of radioactivity.[2,3] The first is that the excised adenoma emits radioactivity at least 20%, and occasionally higher than 50% of postexcision background. Fat, lymph nodes, and even thyroid nodules will never show this level of radioactivity (typical ex vivo emissions less than 3% of background). Ex vivo radioactivity has proved to be 100% accurate in distinguishing parathyroid tissue from fat and lymph nodes when the excised tissue emits 20% of postexcision background radioactivity. This has reduced the number of "diagnostic" frozen sections dramatically with less than 5% of all adenomas falling below this threshold radioactivity level and thereby necessitating a frozen section. We would not recommend eliminating frozen sections until the surgeon has gained significant experience in using the probe in this regard. Because of the systemic administration of the radiopharmaceutical, ex vivo counts must be performed several feet from the patient with the probe aimed away. Operations performed more than 3 hours after sestamibi injection will result in a less radioactive adenoma so this must be kept in mind when measuring ex vivo radioactivity. If in doubt, a frozen section should be performed.

The second important observation is that removal of the radioactive gland will be associated with dramatically decreased gamma emissions within that quadrant of the neck. The loss of this main focus of radioactivity within the neck gives rise to the third observation: the radioactivity in all four "quadrants" of the neck will equalize.[2,3]

The use of the probe allows the resection to be performed rather quickly. Adenomas are identified an average of 11 minutes after incision. We have had five patients with completely intrathyroidal parathyroids each of which were found within 30 minutes of making the skin incision. Their hidden position was disclosed by a decline in radioactivity behind the thyroid and the demonstration of emissions several thousand per-second higher in one specific portion of the thyroid compared to the remainder of the gland. Our average operative time for adenoma resection is now under 30 minutes. Although the speed at which an operation is performed is not important in itself, we believe these times are a reflection of the simplicity of this technique.

Detailed monitoring of the potential radiation hazards has shown this procedure to pose no significant risk to operating room personnel, surgeon, or pathologist.[2] The surgeon's exposure is relatively insignificant and we have reported that the cumulative radiation dose acquired over 15 cases is less than 1% of acceptable yearly exposure (5 rem) as determined by the Nuclear Regulatory Commission. (Radiation safety issues are discussed more fully in another chapter). Similarly, the radioactive adenoma sent to pathology contains only slightly more radioactivity than background (0.04 mR/hr) and therefore poses no exposure hazard to frozen section personnel and does not contaminate the cryostat or other instruments/liquids. The soiled linens and sponges do not require special handling and can be discarded as routine.

SUMMARY

When selected appropriately, a majority of patients with primary HPTH may be successfully treated through a minimally invasive technique. The use of local anesthesia and the limited scope of the dissection afforded by intraoperative nuclear mapping may decrease the incidence of failed explorations and other potential complications associated with this operation. Although intraoperative nuclear mapping can be a very useful tool to minimize the efforts necessary to localize a diseased parathyroid, prior experience in parathyroid surgery is still mandatory since clinical judgment will continue to play the dominant role in determining when the operation is complete and when a bilateral exploration is required.

REFERENCES
1. Denham D, Norman J. Cost-effectiveness of preoperative sestamibi scan for primary hyperparathyroidism is dependent solely upon surgeon's choice of operative procedure. J Am Coll Surg 1998; 186:293-304.
2. Norman J, Chheda H. Minimally invasive parathyroidectomy facilitated by intraoperative nuclear mapping. Surgery 1997; 122:998-1004.
3. Norman J, Denham D. Minimally invasive radioguided parathyroidectomy in the reoperative neck. Surgery 1998; in press.
4. Norman J. The technique of intraoperative nuclear mapping to facilitate minimally invasive parathyroidectomy. Cancer Control, 1997; 4:500-504.

Emerging Applications in Other Skin Cancers

C. Wayne Cruse

INTRODUCTION

The use of lymphatic mapping and sentinel lymphadenectomy has at many institutions profoundly changed the management of patients with malignant melanoma. The number of patients with melanoma is increasing and the death rate for patients with non-thin lesions is significant. This technique allows identification of regional nodal disease at an early stage.

Non-melanoma skin cancer far surpasses the incidence of malignant melanoma, and is the leading cause of cancer in the United States. More than 600,000 cases occur annually. Basal cell carcinoma comprises the majority of these and only very rarely involves the regional lymph nodes. Squamous cell carcinoma (SCC) comprises approximately 20% and metastases to regional lymph nodes gravely influence the prognosis. Lymphatic mapping and sentinel lymphadenectomy may be useful to identify the patients with micrometastatic SCC in the regional lymph nodes. Merkel cell carcinoma (MCC) is an uncommon skin malignancy, has a poor prognosis, and a significant chance for regional lymph node metastases. These patients may also benefit from lymphatic mapping and sentinel lymphadenectomy. Overall, any patient with any skin cancer with a significant chance of lymphatic spread should be considered for this technique.

SQUAMOUS CELL CARCINOMA

Squamous cell carcinoma accounts for approximately 20% of all skin cancers. The incidence in the United States is 41.4 cases per 100,000 and is increasing.[1] The most powerful predictor of survival in patients with SCC is the presence or absence of metastatic disease. Approximately 80% of all metastatic disease initially occurs in the regional lymph nodes. The overall incidence of regional lymph node

Radioguided Surgery, edited by Eric D. Whitman and Douglas Reintgen.
© 1999 Landes Bioscience

metastases in SCC is relatively small, reported between 0.5% and 5%[2] but carries a poor prognosis. Regional metastatic SCC is associated with a 5-year survival rate of 34% and a 10-year survival rate of 16%.[3]

The use of lymphatic mapping and sentinel lymphadenectomy will be beneficial only to the patients at significant risk of lymph node disease. Factors influencing SCC metastatic rate include histological parameters, anatomic site, prior treatment, diameter, immunosuppression, and etiology.

Histological factors are important in predicting metastatic potential as a poorly differentiated histological pattern carries a 3-fold chance of metastases over a well-differentiated pattern. Poorly differentiated lesions comprise only 19.6% of all cutaneous SCC, but account for 51% of all metastasizing lesions.[4] The histologic classification is also important. Adenoid SCC, a histological variant, exhibits a metastatic rate up to 67% in extensive lesions, whereas verrucous SCC has a very low rate of metastases.[5]

Histological depth of invasion according to Clark has been investigated in SCC. The risk of recurrence and metastasis increases greatly with invasion to or below the reticular dermis, which is Clark level IV and V, respectively. Tumor thickness has been noted to be important in determining metastases. In a review, all cases that metastasized were at least 4 mm thick and all deaths were in lesions over 10 mm thick.[6]

The presence of an inflammatory response with a lymphocytic infiltration surrounding the lesion has been noted to diminish the chance for metastases. Histological evidence of perineural invasion occurs in approximately 5% of patients and studies have shown an increased metastatic rate and poor prognosis.[2]

Location of the tumor is important as tumors of the scalp, temple, ears, nostrils and extremities have been reported to be especially prone to metastases. The overall metastatic rate for the ear is 11% and lip is 13.7%.[4]

Prior treatment and subsequent local recurrence of SCC carries a worse prognosis. Recurrent tumors overall have a 25% metastatic rate and recurrent tumors of the ear carry a 45% metastatic rate.[2]

Increased size of the primary tumor is associated with a decreased overall survival and increased rate of regional metastasis. For lesions over 2 cm. in diameter, the metastatic rate is 23.4%. The overall 5-year survival rate for lesions less than 2 cm is 98% and is 72% for lesions over 2 cm.[2]

Patients with an altered immune status are also at higher risk for cancer in general and specifically SCC. Organ transplant patients on immunosuppression have an increased incidence of SCC which increases with time after transplantation. Patients with some lymphoproliferative disorders develop SCC that behaves aggressively.

The etiology influences the frequency of metastases. Carcinomas arising from chronic lesions have a relatively high metastatic rate, generally more than 20%, and are associated with a poor prognosis. These include squamous cell carcinomas arising in burn scars, radiation ulceration, chronic ulcers, osteomyelitis sinuses, traumatic wounds, fistulas, and calluses.[7]

Patients with palpable lymph nodes are treated with a regional lymph node resection. Patients without clinically suspicious regional disease are generally treated without prophylactic regional node resection, but are monitored closely. If regional disease develops clinically later, then it is treated with regional lymph node resection.

Ideally, the above risk factors could identify SCC patients who would benefit from SLN biopsy. Unfortunately, no long term studies with large numbers of patients are available to provide specific information as to which patients would benefit from lymphatic mapping and selective lymphadenectomy. For selected patients at-risk we use lymphatic mapping to identify occult micrometastatic regional nodal SCC. If the sentinel lymph node is negative then no further treatment is necessary. Patients with micrometastatic SCC in the SLN undergo completion node dissection. Some patients may not be candidates for further surgery, and radiation therapy may be considered for treatment of the micrometastatic SCC of the lymph nodes. For example, radiation therapy has been used extensively to treat micrometastatic SCC to the lymph nodes of the neck from primary lesions.

Early experience with SLN biopsy exists in patients wtih SCC of the penis, even predating SLN techniques for melanoma or breast cancer. The "sentinel" lymph node was identified by preoperative lymphangiogram of the penis.[8] Penile cancer spreads in a predictable pattern with inguinal nodes as the primary site of metastases, and bilateral involvement is not uncommon.

If invasion of the shaft or corpora cavernosa occurs, then 33% of patients with clinically negative nodes develop inguinal metastases.[9] The pathological status of the regional lymph nodes is an independent prognostic factor in predicting survival. In addition, the extent of nodal involvement is itself of prognostic importance. Solitary or unilateral nodal metastases are associated with improved survival, relative to bilateral or extensive regional disease.[10] The identification of patients with micrometastatic squamous cell carcinoma in the sentinel lymph node and subsequent early treatment of the inguinal nodal basin may improve survival.

Because the lymphatic drainage of the penis is ambiguous, preoperative lymphoscintigraphy is essential to identify the site(s) of lymph nodes at highest risk for metastases. Intraoperative SLN mapping proceeds as with melanoma, avoiding the high morbidity associated with inguinal node dissection.

Lymphatic mapping and sentinel lymphadenectomy has been successfully used to identify micrometastatic regional squamous cell carcinoma in a patient with an extensive squamous cell carcinoma of the upper extremity. The micrometastatic regional disease was subsequently treated with regional node dissection and the patient had no regional recurrence (Figs. 12.1 and 12.2).[11]

Squamous cell carcinoma of the lip is the most common malignancy of the oral cavity. The incidence of lymphatic involvement is related to the size and location of the primary and to a poor histological grade of differentiation. The cure rate is overall very good, but regional metastases occurs in 2-15% of patients and the 5-year survival rate is dramatically decreased with regional spread of the cancer.

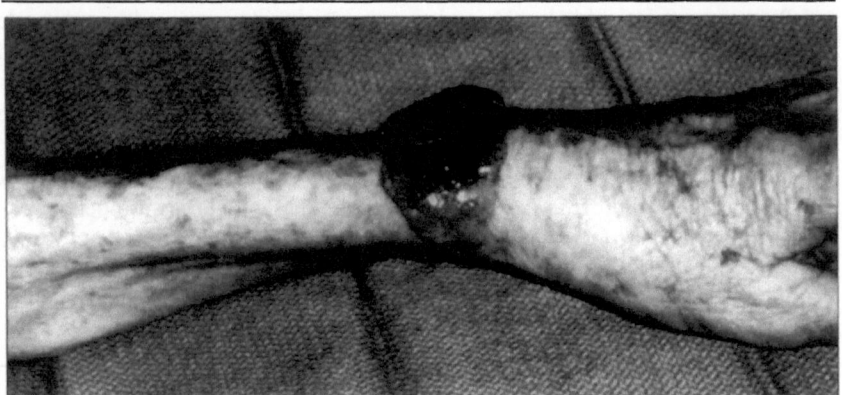

Fig.12.1. 72 year old female with 4 x 3 cm SCC of left wrist involving radius

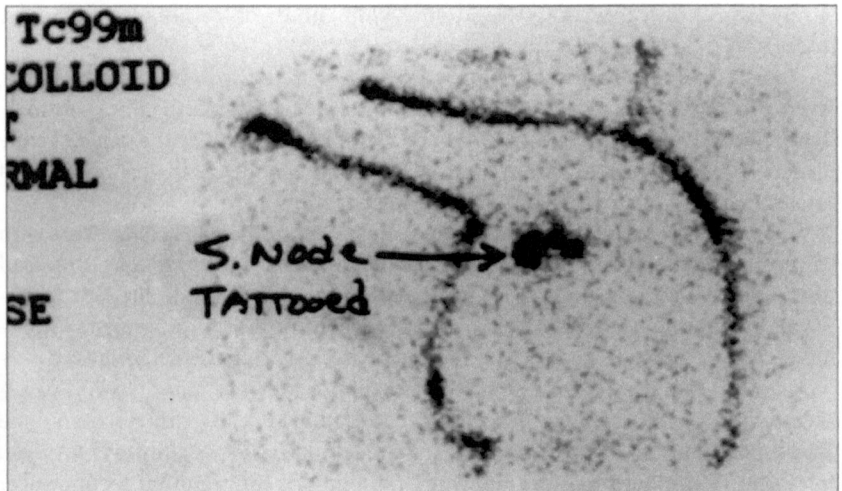

Fig. 12.2. Preoperative radiolymphoscintigram demonstrating two sentinel lymph nodes.

Some have advocated bilateral suprahyoid neck dissections for at-risk patients, then proceed with a complete neck dissection if positive.[12,13]

In patients with lesions over 2 cm, poorly differentiated histologically, and located on the upper lip or at the commissures, the chance for metastases is significant. These patients may benefit from lymphatic mapping and selective lymphadenectomy. In the excision and reconstruction of large lip lesions, extensive face and neck incisions and dissection is necessary, and any lymph nodes located by the mapping techniques could be harvested through incisions used for the excision and reconstruction. The patients that have micrometastatic disease would therefore be identified earlier in the course of disease. The death of patients with squamous cell carcinoma of the lip is usually from uncontrolled local and regional disease and earlier treatment may improve survival.

MERKEL CELL CARCINOMA (MCC)

Merkel cell carcinoma of the skin is an uncommon skin malignancy derived from primitive epithelial cells. The cell was first described by Merkel in 1875[14] and the tumor was first described by Toker in 1972.[15] It has been called neuroendocrine carcinoma of the skin, cutaneous APUDoma, and anaplastic carcinoma of the skin. The Merkel cell is located near the basal layer of the epidermis and functions as a receptor of mechanical stimuli. Clusters of cells form sensitive mechanoreceptors and are found throughout the skin.

Since 1972, over 600 cases have been reported. Patients are predominately over 65, and there are more males than females. The tumor frequently arises in sun-exposed areas such as the head and neck. It begins usually as a smooth dermal nodule, and may become ulcerated. Merkel cell carcinoma is thought by some to be increasing in frequency.

The tumor carries an overall poor prognosis. Local recurrence rates are reported between 26-44%. Regional nodal metastatic disease is present in 50-75% of patients sometime in the course of the disease. Between 7-31% of patients present with clinically evident regional lymph node disease, particularly tumors of the head and neck. Systemic disease is usually preceded by regional lymph node disease.[16]

There is little data on long-term follow-up with adequate numbers of patients. Some studies have shown a 50% regional recurrence rate at 2 years and between 30-55% long term survival rate.

Treatment of the primary tumor consists generally of surgical excision with a 2-3 cm margin. For patients with clinically suspicious regional lymph node disease, a therapeutic lymph node dissection is recommended. In patients with regional metastases, the prognoses is poor.

For patients without clinically suspicious regional disease, the role of prophylactic lymph node resection remains controversial. There are no reliable primary tumor characteristics to predict lymph node metastatic disease. The most important predictor of survival has been the presence of lymph node metastases, and because of the poor prognosis, some authors have recommended prophylactic lymph dissection in all patients.[17]

A hypothetical alternative to prophylactic node dissection is selective lymphadenectomy for all patients with Merkel cell carcinoma.[18] Routine application of SLN mapping to this patient group eliminates the morbidity of the more extensive surgical procedure, while identifying those patients with regional metastases who may benefit from complete node dissection.

We have reported lymphatic mapping and selective lymphadenectomy in 12 patients with Merkel cell carcinoma.[19] Two patients were found to have

12

Fig. 12.3. (See Color Insert for color representation.) 69 year old female with 4 x 5 cm MCC of nasal dorsum.

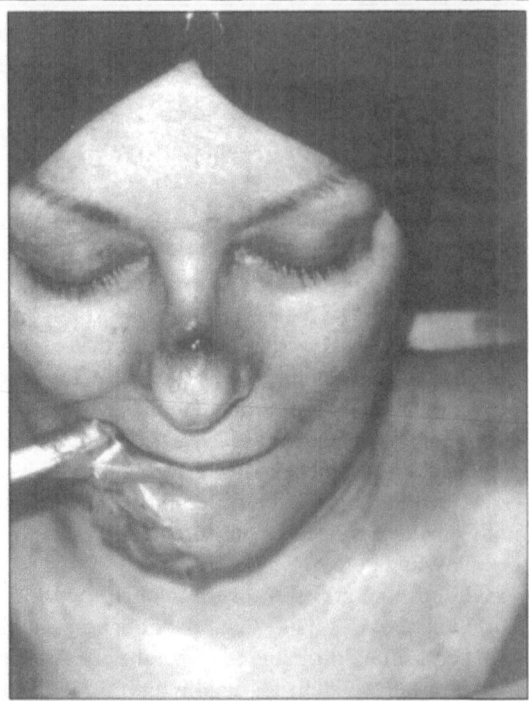

12

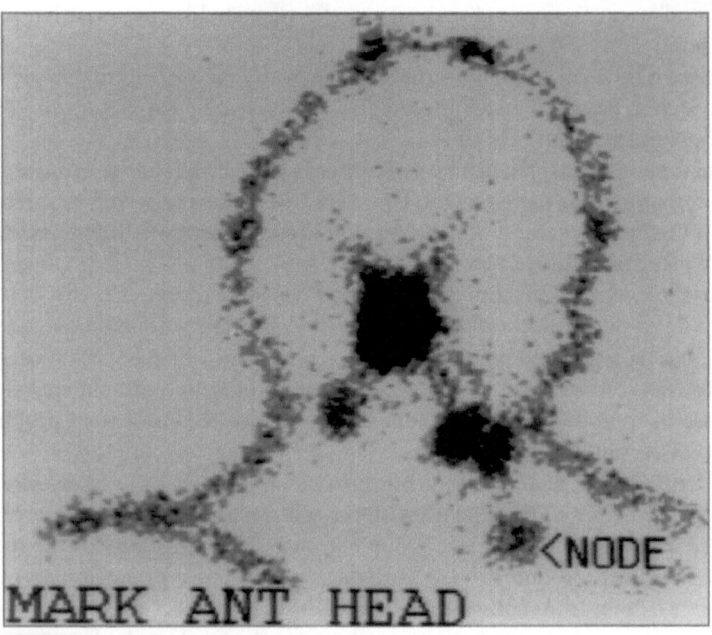

Fig. 12.4. Preoperative radiolymphoscintigram demonstrating sentinel lymph nodes in bilateral submandibular areas.

micrometastatic disease in the sentinel nodes, and complete dissection of the involved nodal basin revealed additional positive nodes. Neither patient has developed regional recurrence. For patients with negative sentinel nodes, no patient developed regional disease at a median follow-up time of 10.5 months. Figures 12.3 and 12.4 illustrate a patient with Merkel cell carcinoma of the nose treated with lymphatic mapping and selective lymphadenectomy. Bilateral cervical sentinel nodes were positive, she subsequently underwent bilateral modified radical neck dissections and is now disease-free at 2 years.[20]

Lymphatic mapping and selective lymphadenectomy has been used in only a few patients with skin cancer other than malignant melanoma. However the anatomical and physiological parameters that make the technique effective in melanoma are present for all skin cancers. No long term data is available in large numbers of patients with other skin cancers, but patients at high-risk for developing regional lymph node disease may benefit from the procedure. The morbidity is relatively low and much benefit may be obtained.

TECHNICAL CONSIDERATIONS

The techniques of lymphatic mapping and selective lymphadenectomy for skin cancers other than malignant melanoma are much the same as those used for malignant melanoma. Preoperative lymphoscintigraphy and intraoperative mapping and biopsy are performed as described in multiple references and elsewhere in this handbook.[21-23]

The head and neck is the predominate site for Merkel cell carcinoma and squamous cell carcinoma. When performing lymphatic mapping in the head and neck, a few specific points should be noted. All incisions for lymphatic mapping and selective lymphadenectomy should be made with considerations for the possibility of subsequent radical neck surgery. Upper neck incisions should be placed transversely to correspond with a later upper neck transverse incision for radical neck dissection. Preauricular incisions can be used liberally to expose the parotid gland and lymph nodes in the area. The sentinel lymph node(s) are found in the usual Level I-V locations and much care must be taken to preserve the many neck structures at risk. Small initial incisions may need to be extended for adequate exposure if difficulty arises in finding the sentinel lymph node(s).

In the head and neck, it is often necessary to excise the primary tumor site in order to perform lymphatic mapping. The radioactivity from the primary tumor site may be very close to the lymph node basin and interfere with intraoperative localization with the hand-held gamma probe. Removing the primary tumor site eliminates this proximity problem.

Primary tumor site excision may require rotation flaps or other reconstructive coverage procedures. Planning for the reconstruction of the primary tumor site must be done prior to making an incision for lymphatic mapping, as a poorly placed incision may interfere with subsequent reconstruction needs.

The preoperative lymphoscintigram should be available for examination during surgery. Often the preoperative lymphoscintigram will identify one predominate sentinel lymph node and the overlying skin will be tattooed, but additional lymph nodes may show uptake also. These secondary lymph nodes may also be identified at surgery and removed.

In conclusion, SLN mapping for non-melanoma skin cancers is performed with near identical techniques as for melanoma cases, with additional requirements as a consequence of the predominately head and neck location of these lesions.

SUMMARY

The use of lymphatic mapping and selective lymphadenectomy in patients with skin cancers other than malignant melanoma needs to be validated with larger groups of patients. However, preliminary data suggest that selective lymphadenectomy may be beneficial in those patients with squamous cell carcinoma at higher risk for metastatic disease, based on published histopathologic variables. Sentinel node mapping is also likely to prove useful in all cases of Merkel cell carcinoma, because of the high reported rate of regional metastases in these patients. As with other malignancies, the application of radioguided surgical techniques allows patients with regional metastatic disease to be identified without the morbidity of prophylactic node dissections, reserving node dissection only for those patients with proven regional metastases.

REFERENCES
1.   Silverberg E, Boring CC, Squires TS. Cancer statistics, 1990. CA 1990:40:9-26.
2.   Johnson TM, Rowe DE, Nelson BR et al. Squamous cell carcinoma of the skin excluding lip and oral mucosa). J Am Acad Dermatol 1992; 26:467-84.
3.   Epstein E, Epstein NN, Bragg K et al. Metastases from squamous cell carcinoma of the skin. Arch Dermatol 1968; 97:245-9.
4.   Rowe DE, Carroll RJ, Day CL. Prognostic factors for local recurrence, metastasis, and survival rates in squamous cell carcinoma of the skin, era, and lip. J Am Acad Dermatol 1992; 26:976-90.
5.   Johnson WC, Helwig EB. Adenoid squamous cell carcinoma (adenocanthoma). A clinicopathologic study of 155 patients. Cancer 1966; 19:1639-150.
6.   Friedman H, Friedman HI, Cooper PH et al. Prognostic and therapeutic use of microstaging of cutaneous squamous cell carcinoma of the trunk and extremities. Cancer 1985; 56:1099-105.
7.   Kwa RE, Campana K, Moy RL. Continuing medical education biology of cutaneous squamous cell carcinoma. J Amer Acad Derm 1992; 26:1-26.
8.   Cabanas RM. An approach for the treatment of penile carcinoma. Cancer. 1997; 2:456-466.
9.   Persky L, deKernion J. Carcinoma of the Penis. CA–A Cancer Journal for Clinicians 1986; 36:258-273.
10.   Gillenwater JY, Grayhack JT, Howards SS et al (eds). Adult and Pediatric Urology 3rd Edition, St. Louis: Mosby, 1996:2002-2042.

11. Stadelmann WK, Javaheri S, Cruse CW et al. The use of selective lymphadenectomy in squamous cell carcinoma of the wrist: A case report. The Journal of Hand Surgery. 1997; 22A:726-731.

12. Marshall KA, Edgerton MT. Indications for Neck Dissection in Carcinoma of the Lip. Amer J Surg 1977; 133:216-217.

13. Koc C, Akyol MU, Cekic A et al. Role of suprahyoid neck dissection in the treatment of squamous cell carcinoma of the lower lip. Ann Otol Rhino Laryngol 1997; 106:787-789.

14. Merkel F. Tastazellen und Tastkorperchen bei den Hausthieren und beim Menschen. Arch Mikr Anat 1875; 11:635-52.

15. Toker C. Trabecular carcinoma of the skin. Arch Dermatol 1972; 105:107-110.

16. Haag ML, Glass LF, Fenske NA. Merkel cell carcinoma diagnosis and treatment. Dermatol Surg. 1995; 21:669-683.

17. Victor NS, Morton B, Smith JW. Merkel cell cancer: Is prophylactic lymph node dissection indicated? The American Surgeon 1996; 62:870-882.

18. Cruse CW, Reintgen D, Glass F et al. Neuroendocrine carcinoma of the skin. Cancer Control 1997; 4:346-348.

19. Messina JL, Reintgen D, Cruse CW et al. Selective lymphadenectomy in patients with merkel cell (cutaneous neuroendocrine) carcinoma. Annals of Surgical Oncology. 4(5):389-395.

20. Javaheri S, Cruse CW, Stadelmann WK. Sentinel node excision for the diagnosis of metastatic neuroendocrine carcinoma of the skin: A case report. Annals of Plastic Surgery 1997; 39(3):299-302.

21. Morton DL, Wen D, Wong JH et al. Technical details of intraoperative lymphatic mapping for early stage melanoma. Arch Surg 1992; 127:392-399.

22. Reintgen DS, Rapaport DP, Tanabe KK, Ross M. Lymphatic mapping and sentinel lymphadenectomy. In: Balch CM, Houghton AN, Sober AJ, Soong S eds. Cutaneous Melanoma 3rd Edition. St. Louis: Quality Medical Publishing Inc. 1998:227-244.

23. Norman J, Wells K, Kearney R et al. Identification of lymphatic basins in patients with cutaneous melanoma. Semin Surg Oncol 1993; 9:224-7.

12

Radioguided Surgery and Vulvar Carcinoma

Edward C. Grendys, Jr., James V. Fiorica

INTRODUCTION

Overall, carcinoma of the vulva is a relatively rare malignancy representing approximately 5% of all female genital tract cancers[1] and 1% of all malignancies in women. The incidence has risen slightly over the past 50 years, perhaps because of an overall rise in the average age of women in the U.S.,[2] or an increase in the prevalence of human papillomavirus (HPV) genital tract infection. The reported increase in vulvar carcinoma in situ, a precursor lesion, has nearly doubled from 1.1-2.1 per 100,000 woman–years between 1973 and 1987,[3,4] mostly in younger women.

Fortunately, the majority of lesions are diagnosed at a relatively early stage and are thus amenable to local surgical extirpation with high overall cure rates. Generally, these tumors are readily visible on the external vulvar surface and produce the typical symptom of pruritus in over 90% of patients.[5] Other signs and symptoms associated with vulvar carcinoma include a visible, palpable vulvar mass, pain, bleeding, ulceration, dysuria and/or an abnormal vaginal, perineal discharge. Unfortunately, even given the high prevalence of early symptoms, patient reporting and subsequent diagnostic evaluation is often delayed, most commonly due to patient denial or embarrassment. A histologic diagnosis should always be considered before initiating any topical therapy for women with these complaints.

Vulvar malignancies with the rare exception of vulvar sarcomas appear most commonly in women between the ages of 65 and 75 while the median age of the precursor in situ lesion is approximately 45-50. However, 15% of vulvar carcinomas occur in women less than 40 years of age[6] with multifocal disease being more prevalent.

Radioguided Surgery, edited by Eric D. Whitman and Douglas Reintgen.
© 1999 Landes Bioscience

The vulva is covered by keratinized squamous epithelium and the majority of these neoplasms are of squamous cell (epidermoid) histology (see Table 13.1). Though relatively uncommon, vulvar melanoma represents about 5-10% of malignant vulvar neoplasms and overall about 17% of melanomas diagnosed in females will originate on the vulva. The aggressive biopsy of hyperpigmented vulvar lesions using standard criteria should be considered (See Table 13.2).

EPIDEMIOLOGY

Several infectious agents have been proposed as possible etiologic factors in the development of vulvar carcinoma, including various granulomatous infections, herpes simplex virus and human papillomaviruses (HPV). The molecular role of the human papillomavirus in cervical dysplasia and carcinoma as well as the association between viral induced vulvar condylomata and subsequent development of vulvar carcinoma has been well established.[7] Many investigators have identified HPV DNA in both invasive and precursor carcinoma in situ vulvar lesions.[8] With the increasing incidence of HPV related dysplasias in the younger female population and the reported trend toward a younger age of diagnosis[9] there is certainly cause for concern that an increase in related vulvar cancer may be forthcoming. In a study by Mitchell et al[10] 169 women with invasive vulvar carcinomas were noted to have a secondary genital squamous neoplasm in 13% of

Table 13.1.Histologic distribution of malignant vulvar neoplasms

Tumor type	Percent
Epidermoid	86.2
Melanoma	4.8
Sarcoma	2.2
Basal cell	1.4
Bartholin gland	1.2
Squamous	0.4
Adenocarcinoma	0.6
Undifferentiated	3.9

Plentl, AA, Friedman, EA, Lymphatic system of the Female Genitalia Philadelphia, 1971, WB Saunders

Table 13.2. Indications for excisional biopsy of vulvar nevi

Change in surface area of nevus
Change in lesion contour or surface
Change in lesion color
Change in sensation

13

cases. Brinton et al[11] concluded that women with a history of genital warts, previous abnormal Papanicolaou smears as well as a history of smoking are at increased risk for vulvar cancer. Chronic immunosuppression has also been linked to an increased incidence of both vulvar and cervical disease.[9] Other observational associations have been made between hypertension, diabetes mellitus, and obesity.[12]

ANATOMY AND LYMPHATIC DRAINAGE

The vulva, including the mons pubis, labia majora and minora, clitoris, vaginal vestibule, perineal body and the supporting subcutaneous structures, develops embryologically from the genital tubercle of the cloacal membrane, as does the distal vagina. Because of their embryologic derivation the vulva and distal vagina share common routes of lymphatic drainage. The vulvar lymphatics run anteriorly through the labia majora, turn laterally at the mons pubis and drain primarily into the superficial inguinal lymph nodes. Previous lymphatic dye mapping studies by Parry-Jones[13] demonstrated that the vulvar lymphatic channels do not cross to the contralateral lymph nodes unless the dye is injected in the midline structures (clitoris or perineal body). There may be some minimal, direct drainage to the pelvic lymph nodes though the clinical relevance of these communications appears negligible. Most studies indicate an orderly progression of lymphatic drainage from the vulva to the superficial inguinal lymph nodes then directly to the deep inguinal nodes prior to proceeding to the pelvic nodal structures (See Fig. 13.1).

SURGICAL MANAGEMENT AND STAGING OF VULVAR CARCINOMA

Vulvar cancers have three modes of spread: 1) direct extension into adjacent organs; 2) embolization into locoregional lymph nodes; and 3) hematogenous spread. Fortunately well documented clinical investigations have revealed that early spread is almost always confined to the local inguinal lymphatic basins. Although lymphatic drainage usually proceeds from the superficial to the deep inguinal (femoral) lymph nodes, care must be taken before adopting too cavalier an approach in that deep nodal involvement has been reported without evidence of superficial inguinal lymph node disease.[15-17]

Few radical procedures in gynecologic oncology have been as successful and yet continue to change as much as the surgical approach to vulvar carcinoma. Original radical procedures as described by Taussig[18] and Way[19] defined the procedure that dramatically reduced the mortality from vulvar carcinoma. These traditional "Longhorn" resections included the entire vulva, mons and laterally extending over and including the inguinal lymph nodes and inferiorly to the urogenital diaphragm. Until recently, this procedure was the treatment of choice for all resectable vulvar lesions regardless of clinical size or location. The conventional radical vulvectomy and bilateral inguinofemoral lymphadenectomy had been

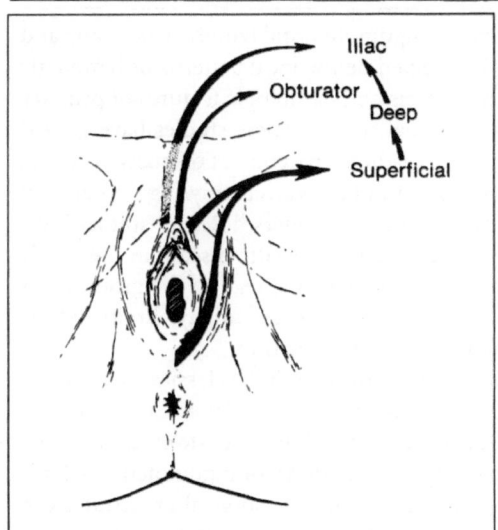

Fig. 13.1. Lymphatic drainage from the external genitalia to the inguinal lymph nodes. Reprinted with permission from DiSai and Creasman, Clinical Gynecologic Oncology, Copyright 1997; 206-207. © 1997 Mosby-Year Book Inc.

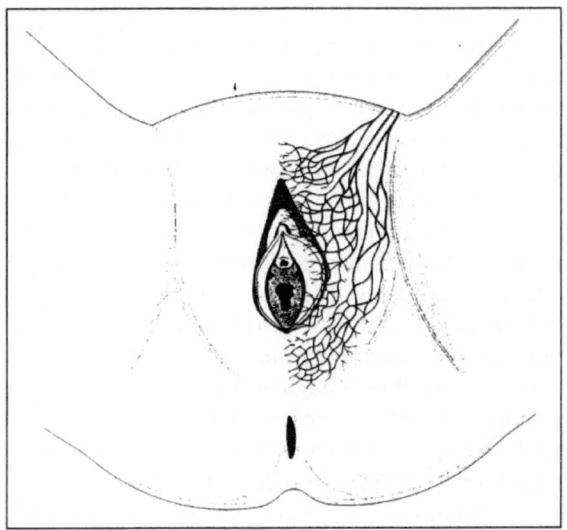

extremely successful from a curative standpoint as has been well documented in the literature.[12,20,21] The overall uncorrected 5-year survival following this procedure is approximately 65% and operability rates in this generally older population about 95%.[22] Unfortunately postoperative morbidity remained extremely substantial. Reported operative and perioperative mortality rates in modern series range from 1-5%.[20-22] The incidence of acute wound breakdown approaches 85%[22] and often disabling chronic lower extremity edema in 30-70%.[23] Genital prolapse, urinary incontinence, and vaginal stricture occurs in 15-20% of women postoperatively with Hacker et al reporting a reoperation rate of 15% for correction of postsurgical complications.[20]

The radical vulvectomy with bilateral inguinofemoral lymphadenectomy and often pelvic lymph node sampling has helped define local patterns of lymphatic dissemination, incidence of nodal metastasis and pathologic features of primary lesions which help predict the risk of nodal spread. These studies have guided gynecologic oncologists in redefining this radical procedure. For many years the addition of a pelvic node dissection to the radical vulvectomy/inguinofemoral lymphadenectomy was standard practice at many centers.[24] Recent studies indicate that fewer than 6% of vulvar cancers have metastatic disease in the pelvic nodes and essentially no patient without evidence of spread to the groin nodes had pelvic lymph node metastasis. This discovery has essentially eliminated the routine use of pelvic lymph node sampling in this situation.

Most gynecologic oncologists have moved away from the classical radical vulvectomy with complete en bloc resection through a single incision. Current trends are moving towards less extensive vulvar surgery with more customization based on lesion size and location, and separate groin incisions for inguinofemoral lymphadenectomy. Based on surgical-pathologic indicators surgical procedures can be uniquely customized based on lesion size, location, and depth of invasion as well as clinical suspicion of lymph node metastasis. Especially true of small (< 2 cm) lateralized lesions with < 1 mm depth of invasion, a wide radical excision with approximately 10 mm free margins (fresh, nonfixed tissue)[25] can be performed with no need to perform a lymphadenectomy thus sparing the patient the potential morbidity. The risk of groin node metastasis as related to depth of invasion, lesion diameter and clinical suspicion and has been well documented by the Gynecologic Oncology Group (GOG)[26] (see Table 13.3) and clearly defines the rationale for more conservative therapies in the early lesion.

The decision to perform inguinal-femoral lymphadenectomy is based upon tumor size, location and depth of invasion as well as clinical suspicion of lymph node metastasis. Recalling that the only patient without risk of nodal spread has tumor invasion of < 1 mm, the decision to perform groin node dissection is relatively straightforward. In general the prevailing thought is that a small, well lateralized lesion can be managed with ipsilateral inguinofemoral lymph node dissection unless a metastatic deposit is noted in which case the contralateral lymphadenectomy should also be performed. Midline lesion resection should include bilateral lymphadenectomies. Controversy remains over whether a superficial and deep groin node dissection should be performed routinely. There is little data to support the notion that superficial dissection alone decreases resultant lymphedema. Metastases to the femoral nodes without involvement of superficial nodes has been reported[15-17] and the Gynecologic Oncology Group study revealed an unexpectedly high incidence of ipsilateral groin recurrences following superficial dissection alone.[27] With the high mortality associated with recurrent groin metastasis, the possible therapeutic benefit of groin node dissection and no documented benefit in limited dissection, our institution continues to perform both superficial and deep lymphadenectomies in patients not placed on study protocol.

INTRAOPERATIVE LYMPHATIC MAPPING
FOR VULVAR CARCINOMA

In 1979 DiSaia et al[28] described the eight to ten inguinal nodes above the crib-riform fascia as "sentinel nodes." The concept of sentinel lymph node identification using intraoperative mapping with lymphoscintigraphy offers the unique potential of decreasing the morbidity associated with inguinal lymphadenectomy. Previous studies have examined lymphatic mapping using lesionally injected technetium[99] sulfur colloid in breast carcinoma[29,30] as well as cutaneous melanoma.[31] These investigators defined the sentinel lymph node as the first node in the lymphatic basin that receives primary lymphatic flow from the suspect lesion and therefore should be the first site of metastatic disease. These and other studies in breast carcinoma and cutaneous melanoma have revolutionized the surgical treatment of these diseases while dramatically reducing associated morbidity, hospital stay and subsequent cost.

Given the basic understanding of vulvar lymphatic drainage investigators have begun to examine the role of selective sentinel lymph node biopsy in early vulvar carcinomas and melanomas. Levenback et al[32] described a method of intradermal injection of isosulfan blue dye which is a triphenylmethane dye transported via lymphatic vessels and concentrated in lymph nodes. Subsequent groin dissection was able to visualize a sentinel node in 7 of 12 samples. In no patient so studied was a positive nonsentinel node identified in the presence of a negative sentinel lymph node. A modification of this approach using technetium[99] sulfur colloid lymphoscintigraphy provides a relatively simple method of intraoperative identification of the sentinel lymph nodes.

TECHNIQUE OF VULVAR LYMPHOSCINTIGRAPHY

After appropriate consent is obtained patients are placed in the dorsal lithotomy position in an outpatient setting approximately 2 hours prior to surgery. No sedation is generally required. The vulvar lesion (or biopsy site) is appropriately identified, prepped and approximately 400 microcuries of technetium[99] sulfur colloid (reconstituted in approximately 8 ml total) is injected intralesionally in a circumferential manner using a 25 gauge spinal needle (see Fig. 13.2). Minimal discomfort is to be anticipated and no local analgesia is usually required and in fact may potentially decrease uptake of the radioactive colloid.

The patient is taken to the operating room about 2 hours after injection, anesthetized, positioned in modified lithotomy position to facilitate inguinofemoral lymph node dissection and prepped and draped in the usual manner. A standard incision is made approximately 2 cm below and parallel to the inguinal ligament. The dissection is carried down through the level of the superficial fascia and a flap is dissected both cephalad and caudad in the usual manner. A intraoperative gamma counter is next brought onto the field. Using the gamma counter the area of greatest activity is identified and careful dissection is used to identify the sentinel lymph

Table 13.3. Risk of lymph node metastasis in relation to depth of tumor invasion

Invasion (mm)	% Ipsilateral node +	%Contralateral only +	% Bilateral +
<2	6.8%	0.0	0.0
3-5	20.4	1.9	2.8
6-10	28.8	3.8	11.3
>11	36.7	2.5	50

Homesley H, Prognostic factors for groin node metastasis in squamous cell carcinoma of the vulva (a Gynecologic Oncology Group Study). Gynecol Oncol 1994; 49:279.

node. After excision it is important to recheck the excised sample to be certain the activity was precisely within the removed lymph node. In our experience, reexamination of the nodal basin can occasionally identify a second or third node with significant radioactive uptake. These nodes should also be excised and identified as sentinel lymph nodes. Given the experimental nature of this procedure, at present we then continue with a complete inguinofemoral lymphadenectomy.

We have published the results of a pilot study of SLN mapping in ten patients with vulvar cancer.[33] Since this was a pilot study, subsequent complete inguinofemoral lymphadenectomy was then undertaken. Of the 10 patients studied, 3 were noted to have metastatic disease all of which were correctly identified in the sentinel node. No patient with a negative sentinel lymph node was found to have metastatic spread to other areas (false negative rate of 0%).

VULVAR MELANOMA

The major treatment strategy for vulvar melanoma remains surgical. As with squamous cell carcinoma there has been a trend away from the complete radical vulvectomy towards a more conservative wide radical resection. Based on work by Trimble et al,[34] a 2 cm margin with excision deep to the level of the inferior urogenital diaphragm appears adequate with similar survival. Inguinal lymph node metastases are rare in patients with Clark's level I or II melanoma[35] and lymphadenectomy can most likely be avoided in these patients. Although regional lymph node dissection in more advanced lesions serves more of a prognostic value rather than a therapeutic one,[36] most investigators still recommend inguinofemoral lymphadenectomy for lesions greater than Clarks level II.[37]

CONCLUSION AND THE FUTURE

In general the role of ultra-radical vulvar resection and inguinofemoral lymphadenectomy has undergone extensive revision in the past 10-15 years with the ultimate goal being to decrease physical as well as psychosocial and psychosexual morbidity. With continuing advances in technology and refinement in surgical

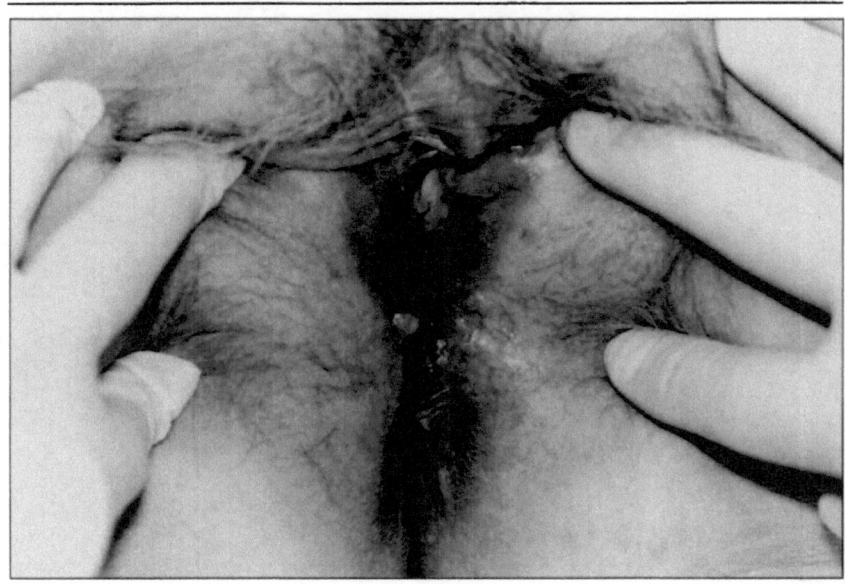

Fig. 13.2a. Posterior squamous cell carcinoma of the vulva.

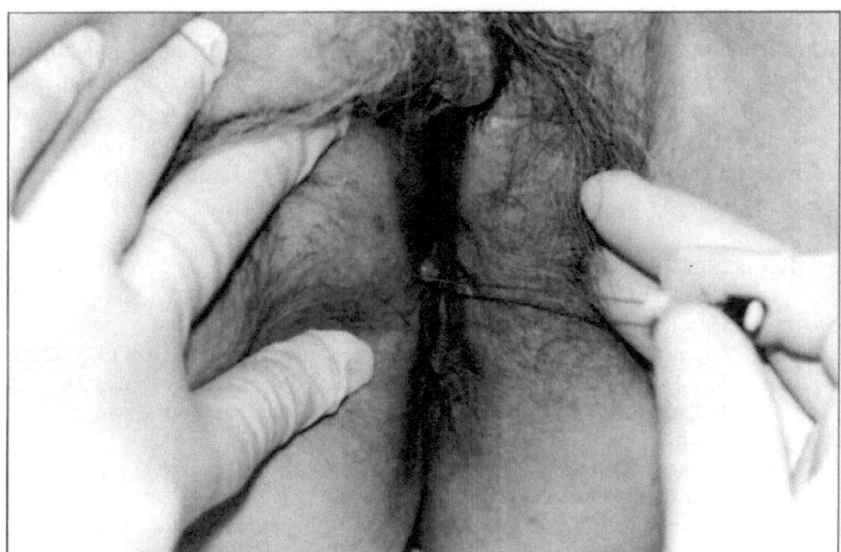

Fig. 13.2b. Insertion of the 25 gauge spinal needle.

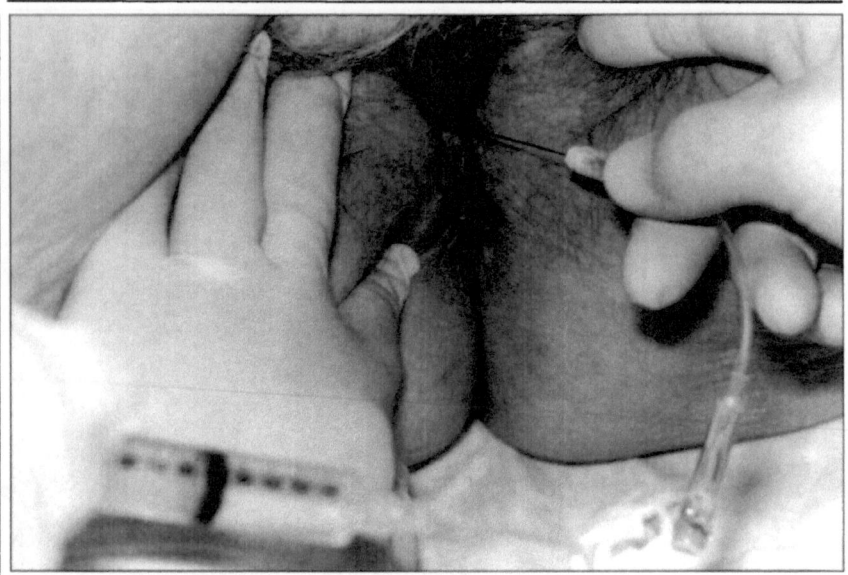

Fig. 32.2c. Injection of technetium[99] sulfur colloid.

techniques it is hoped that sentinel lymph node biopsy will someday replace the complete lymphadenectomy as currently practiced. As with other malignancies, sentinel lymphadenectomy for vulvar carcinoma may refine staging and surgical therapy, eliminating (or at least reducing) the need for prophylactic lymphadenectomy and converting all lymphadenectomies to therapeutic. However, at present, we believe that SLN mapping for patients with vulvar cancer should only be performed under a protocol setting, especially since prior GOG studies suggest a therapeutic benefit to routine complete inguinofemoral lymphadenectomy.

REFERENCES

1. DiSai PH, Creasman WT. Invasive Carcinoma of the Vulva. Clinical Gynecologic Oncology. St. Louis: Mosby-Year Book, Inc., 1997.
2. Green T. Carcinoma of the vulva; A reassessment. Obstet Gynecol 1978; 52:462.
3. Eifel P, Berek J, Thigpen J. Cancer of the cervix, vagina, and vulva. In: Devita V, Hellman S, Rosenberg S eds. Cancer: Principles and Practice of Oncology. Philadelphia: Lippincott-Raven Publishers 1997.
4. Sturgeon S, Brinton L, Devesa S et al. In situ and invasive vulvar cancer incidence trends. Am J Obstet Gynecol 1992; 166:1482.
5. Burke T, Eifel P, McGuire W et al. Vulvar Carcinoma. In: Hoskins W, Perez C, Young R eds. Principles and Practice of Gynecologic Oncology. Philadelphia: Lippincott-Raven Publishers 1997.
6. Rutledge F, Mitchell M, Munsell M et al. Prognostic indicators for invasive carcinoma of the vulva. Gynecol Oncol 1991; 47:239.
7. Brinton L, Nasca P, Mallin K et al. Case-controlled study of cancer of the vulva. Obstet Gynecol 1990; 75.

13

8. Downey G, Okagaki T, Ostrow R et al. Condylomatous carcinoma of the vulva with special reference to human papillomavirus DNA. Obstet Gynecol 1988; 72:68.

9. Carter J, Carlson J, Fowler J et al. Invasive vulvar tumors in young women: A disease of the immunosuppressed? Gynecol Oncol 1993; 51:307.

10. Mitchell M, Prasad C, Silva E et al. Second genital primary squamous neoplasms in vulvar carcinoma: Viral and histopathologic correlates. Obstet Gynecol 1993; 81:13.

11. Brinton L, PC N, Mallin K et al. Case-controlled study of cancer of the vulva. Obstet Gynecol 1990; 65:859.

12. Franklin E, Rutledge F. Epidemiology of epidermoid carcinoma of the vulva. Obstet Gynecol 1972; 39;165.

13. Parry-Jones E. Lymphatics of the vulva. J Obstet Gynecol Br Empire 1963; 70:751.

14. Shimm D, Fuller A, Orlow E et al. Prognostic variables in the treatment of squamous cell carcinoma of the vulva. Gynecol Oncol 1986; 24:343.

15. Parker R, Duncan I, Rampone J et al. Operative management of early invasive epidermoid carcinoma of the vulva. Am J Obstet Gynecol 1975; 123:349.

16. Hacker N, Nieberg R, Berek J et al. Superficially invasive vulvar cancer with nodal metastases. Gynecol Oncol 1983; 15:65.

17. Chu J, Tamimi H, Figge D. Femoral node metastases with negative superficial inguinal nodes in early vulvar cancer. Am J Obstet Gynecol 1981; 140:337.

18. Taussig F. Carcinoma to the vulva: An analysis of 155 cases. Am J Obstet Gynecol 1970; 40:764.

19. Way S. Carcinoma of the vulva. Am J Obstet Gynecol 1960; 79:692.

20. Hacker N, Leuchter R, Berek J et al. Radical vulvectomy and bilateral inguinal lymphadenectomy through separate groin incisions. Obstet Gynecol 1981; 58:574.

21. Iverson T Aalders J, Christensen A et al. Squamous cell carcinoma of the vulva: A review of 424 patients, 1957-1974. Gynecol Oncol 1980; 9:271.

22. Podratz K, Symmonds R, Taylor W. Carcinoma of the vulva: Analysis of treatment failures. Am J Obstet Gynecol 1982; 143:340.

23. Andreasson B, Bock J, Weberg E. Invasive cancer in the vulvar region. Acta Obstet Gynecol Scand 1982; 61:113.

24. Monaghan J, Hamond I. Pelvic node dissection in the treatment of vulvar carcinoma, is it necessary? Br J Obstet Gynecol 1984; 91:270.

25. Heaps J, Fu Y, Montz R et al. Surgical-pathologic variables predictive of local recurrence in squamous cell carcinoma of the vulva. Gynecol Oncol 1990; 38:309.

26. Homesley H. Prognostic factors for groin node metastasis in squamous cell carcinoma of the vulva (a Gynecologic Oncology Group Study). Gyencol Oncol 1994; 49:279.

27. Stehman F, Bundy B, Dvoretsky P et al. Early stage I carcinoma of the vulva treated with ipsilateral superficial inguinal lymphadenectomy and modified radical hemivulvectomy: A prospective study of the Gynecolgic Oncolgy Group. Obstet Gynecol 1992; 79:490.

28. DiSaia P, Creasman W, Rich W. An alternate approach to early cancer of the vulva. Am J Obstet Gynecol 1979; 133:825.

29. Drag D, Weaver D, Alex J et al. Surgical resection and raidolocalization of the sentinel lymph node in breast cancer using a gamma probe. Surg Oncol 1993; 2:335.

30. Guiliano A, Kirgan D, Guenther J et al. Lymphatic mapping and sentinel lymphadenectomy for breast cancer. Ann Surg 1994; 220:391.

13

31.   Morton D, Wen D, Wong J et al. Technical details of intraoperative lymphatic mapping for early stage melanoma. Arch Surg 1992; 127:392.

32.   Levenback C, Burke T, Gershenson D et al. Intraoperative lymphatic mapping for vulvar cancer. Obstet Gynecol 1994; 84:163.

33.   DeCsare S, Fiorica JV, Roberts W et al. A pilot study utilizing intraoperative lymphoscintigraphy for identification of the sentinel lymph nodes in vulvar cancer. Gynecol Oncol 1997; 66:425.

34.   Trimble E, Lewin J, Williams L. Management of vulvar melanoma. Gynecol Oncol 1992; 45:254.

35.   Clark W, Bernardino E, From L et al. This histogenesis and biologic behavior of primary human malignant melanomas of the skin. Cancer Res 1969; 29:705.

36.   Phillips G, Bundy B, Okagaki T et al. Malignant melanoma of the vulva treated by radical hemivulvectomy: A prospective study of the Gynecologic Oncology Group. Cancer 1994; 73:2626.

37.   Homesley H. Management of vulvar cancer. Cancer 1995; 76:2159.

13

Bone Lesion Localization

Lary A. Robinson

INTRODUCTION

Accurate staging plays a critical role in the evaluation of any malignancy because it has a pivotal role in determining therapy. Since stage and survival are strongly correlated, the prognosis is also determined by staging. Metastases to bone from any solid tumor are classified as distant, blood-born metastases and the tumor is considered Stage IV.[1] Documented bone metastases eliminates surgery as a curative option, generally mandating a chemotherapy approach occasionally with radiotherapy added for symptomatic lesions.

In one study in 1976, as many as 15% of primary extraosseous malignancies have solitary abnormalities on radioisotope bone scan,[2] yet from 36%[2]-71%[3] of these bone lesions were benign on biopsy. With modern, more sensitive imaging equipment, it is likely that more routine bone scans will have at least one abnormal focus leading to an even higher false positive rate of suspected bone metastasis. This high false positive rate in solitary bone scan abnormalities could potentially lead to over-staging of a malignancy. Therefore, it is imperative that there is histologic confirmation of suspected osseous metastases, especially when they are discovered only on radioisotopic bone scans.

BONE METASTASES

Despite the fact that primary bone malignancies are rare, metastases to the bone are relatively common and tend to be found in sites of persistent red marrow, particularly the axial skeleton.[4] The most common sites of metastases in a series of 2001 patients with known bone metastases was (in decreasing frequency) the vertebrae, pelvis and sacrum, femur, ribs, skull, humerus, scapula, and sternum.[5]

Metastases to bone most commonly are seen in breast and in prostate carcinomas. Kidney, thyroid and lung cancer bony metastases occur less frequently.[4] Cancers of the gastrointestinal tract, sarcomas, and genital tract rarely metastasize

Radioguided Surgery, edited by Eric D. Whitman and Douglas Reintgen.
© 1999 Landes Bioscience

to bone. The overall frequency of osseous metastases is 67% in breast cancer patients, 50% in prostate cancer, 25% for lung cancer and kidney cancer, and less than 10% in all other malignancies.[4]

Most bony metastases present with symptoms, usually new onset bone pain and occasionally swelling over the metastatic site. Rarely, a pathologic fracture is the first sign of metastatic disease to the bone. An elevation of the serum alkaline phosphatase may suggest the presence of bony metastases, but this enzyme may be elevated due to many other causes such as hyperparathyroidism, osteomalacia, osteitis deformans, osteogenic sarcoma, rickets, pregnancy, healing fractures, normal growth, and a variety of hepatobiliary conditions.[6] The finding of an elevated serum calcium is even more uncommon and less specific as an indicator of bony metastases.

Clinically-silent bone metastases are infrequent and are suspected from an abnormal staging radioisotopic bone scan. For some malignancies, a bone scan is routine in the initial workup and asymptomatic abnormalities may be found. For other cancers, such as nonsmall cell lung cancer, a bone scan is only recommended if there is some definite clinical indicator such as new onset bone pain or an elevated serum calcium or serum alkaline phosphatase. Osseous metastases are rarely found in nonsmall cell lung cancer in the absence of definite clinical indicators.[7,8]

The usual plain radiographic finding suggestive of bony metastases is decreased density at the site of the metastasis. Most bone metastases cause bone destruction giving this osteolytic picture. Much less common are osteoblastic or osteosclerotic changes from increased bone density that occasionally is caused by prostate cancer or from hormone treatment in metastatic breast cancer.[5] However, the plain bone radiograph is a relatively insensitive indicator of metastatic disease since at least 50% of trabecular bone must be destroyed before it is radiographically visible.

RADIOISOTOPE BONE SCAN

A much more sensitive indicator of metastatic disease is the radioisotopic bone scan, which is generally positive when as little as 5-15% of trabecular bone is destroyed by tumor.[4] Therefore, the bone scan may well demonstrate a bony metastasis much earlier than a plain bone radiograph. Indeed, in only 3% of patients will bone metastases be visible on a plain radiograph when the bone scan is normal.

The mechanism allowing for uptake of the radiopharmaceutical, usually ^{99m}technetium-labeled diphosphonate, on a bone scan is incompletely understood. Areas of increased bone formation and/or increased blood flow localize the radioisotope[9] to cause increased uptake ("hot spot") on bone scan. Metastatic tumor will cause local bone destruction but also there will generally be synchronous bone formation. The radiopharmaceutical appears to bind to the surface of the hydroxyapatite crystal of the newly forming bone, and not to the tumor cells themselves.

A focal decrease in the uptake of the radioisotope, a "photopenic lesion," is unusual and is generally seen with areas of decreased blood flow or rarely a metastatic tumor with decreased or absent new bone formation.[9] A rapidly-growing, destructive metastatic tumor may cause bone resorption before new bone mineralization occurs. Therefore, the radioisotope is not taken up in this localized area of metastatic tumor. However, these instances are rare and most bone scan lesions demonstrate increased uptake.

Despite the high sensitivity of the bone scan, it is not particularly specific in demonstrating metastatic tumor in the bone. The bone scan may have localized areas of increased uptake of radioisotope from a wide variety of normal or benign causes listed in Table 14.1, which may lead to a false-positive result in the staging of the primary malignancy. The nonspecific nature of a positive bone scan, especially when plain radiographs are normal, mandates biopsy and histologic confirmation of the suspected metastatic tumor to definitively stage the primary malignancy and determine subsequent treatment.

TECHNIQUES OF BONE LESION LOCALIZATION

Metastatic disease to the bone may be readily apparent in some patients who have localizing symptoms. When local bone pain or swelling is present, plain radiographs often will demonstrate the lesion. A bone scan may then be obtained if these plain radiographs are equivocal, but also a bone scan may be helpful to search for other metastatic sites particularly on weight-bearing bones such as the femur, so that they can be treated before there is a pathologic fracture. Other radiological techniques such as computed tomography (CT) or particularly magnetic resonance imaging (MRI) with gadolinium contrast appear to be quite sensitive, especially with the spine and pelvis, to locate metastases when the plain radiographs are normal and the bone scan is abnormal.[4] An MRI of the spine may provide important information in patients with neurologic symptoms or vertebral body collapse. The MRI will delineate the extent of the tumor mass to determine if spinal cord impingement is imminent or present, and to allow planning for possible neurosurgical intervention. However, stepping straight to an initial "screening" MRI of the spine in the patient with advanced cancer and new back pain to locate and delineate the extent of metastatic disease to the spine may be more efficient and cost-effective than the conventional approach of first obtaining the plain radiographs, the bone scan, and then the spine MRI.[10]

In the patient with a soft-tissue mass or a large lytic lesion in the bone, a percutaneous needle biopsy usually will provide histologic confirmation of the metastatic tumor. A CT-directed needle biopsy of a spine lesion is also often successful. However, because of the potential for sampling error when a needle biopsy is nondiagnostic, a open surgical biopsy may be advisable when staging and treatment depends on obtaining a malignant histologic result. With local bone symptoms and a corresponding plain radiographic lesion, intraoperative localization of the bone abnormality for surgical biopsy is usually not difficult, especially if the target bone is a rib and the patient is thin. Nevertheless, in the obese or very muscular patient, it still may be difficult to find the correct rib without a large incision and multiple intraoperative radiographs or fluoroscopy.

14

Table 14.1. Conditions causing increased uptake of tracer on bone scan*

•Normal structures:
  Sternomanubrial and corpus-manubrial joints
  Base of skull, facial bones, inferior tip of scapula
  Alae of sacrum
  Kidneys and bladder
  Variant anatomy
•Increased blood flow and bone formation:
  Sudeck's atrophy
  Hyperostosis frontalis interna
  Osteitis pubis
  Renal osteodystrophy
  Eosinophilic granuloma
  Fibrous dysplasia
  Paget's disease
  Aseptic necrosis, cysts
  Osteoid osteoma
•Bony abnormalities with increased blood flow:
  Osteomyelitis, osteitis
  Fracture (recent or healing)
•Soft-tissue abnormalities:
  Calcific tendinitis or myositis
  Hydronephrosis or hydroureter
  Dental abscess
  Injection site
  Postoperative scar
  Soft tissue osseous metaplasia
•Benign bone tumors:
  Fibroma
  Chondroma
•Primary malignant bone tumors:
  Chondrosarcoma
  Osteosarcoma
  Ewing's sarcoma
•Metastatic tumors

---

*Adapted from Wahner HW and Brown ML,[9] with permission.

When there is a bone scan abnormality, especially if there is more than one, and the plain radiographs are normal in an asymptomatic patient with a known cancer, most radiologists feel that this indicates metastatic disease in bone.[2] Figure 14.1 illustrates a typical patient with a positive bone scan with suspected metastatic cancer but the plain bone radiographs were normal and the patient (quite obese) had no localizing symptoms. However, it appears that this commonly-held belief by radiologists is incorrect since there is a bone biopsy-documented, very high incidence of false-positive bone scan results in this setting, ranging from 47-71% in various studies.[3,8,11,12] That is, benign lesions frequently account for these bone scan abnormalities and a confirmatory open biopsy is mandatory. Still, there are increased technical problems for the surgeon to find the exact lesion in the bone when only the bone scan serves as a guide.

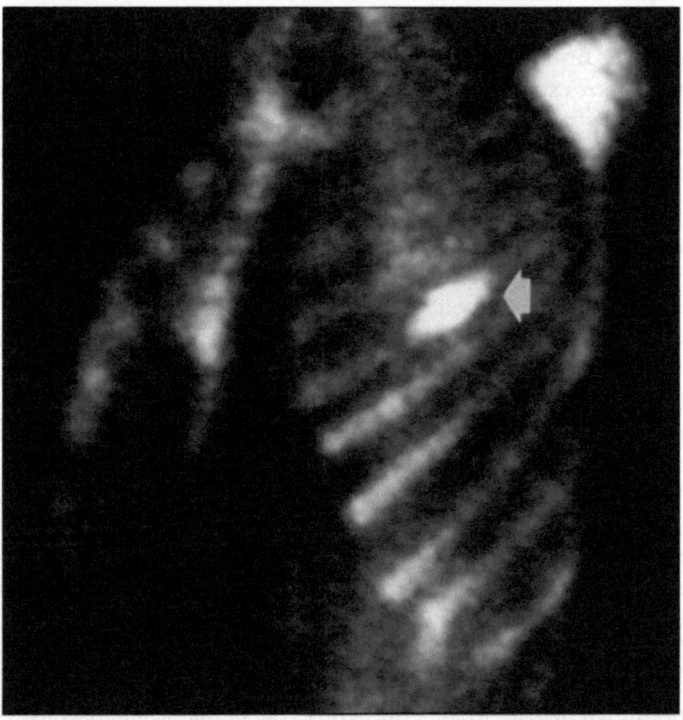

Fig. 14.1. Left antero-oblique view of a radioisotope bone scan (coned down view of the thorax only) of a patient with suspected carcinoma metastatic to bone. An area of increased tracer activity (white arrow) is seen in the anterolateral aspect of the left 5th rib. The rib detail radiographs of this area of the 5th rib were normal. This small area of the rib was biopsied using gamma probe guidance and it proved histologically to be a benign chondroma.

CONVENTIONAL SURGICAL BIOPSY

The surgeon faces a difficult clinical problem when asked to do a biopsy of an area of increased uptake on bone scan in an asymptomatic patient with normal plain bone radiographs. When positive on bone scan, ribs are technically the easiest to biopsy but localization of the exact site is challenging. Depending upon the quality and definition of the image on the bone scan, it may be difficult just to define exactly which rib is "hot" and even more challenging to localize the precise anterior-posterior location of the lesion.

In this setting, the surgeon may be forced to make a substantial incision to excise a large area of one or even two ribs which are usually grossly normal in order to be sure that the abnormality is included in the specimen. Equally frustrating is the difficulty at surgery of actually counting which is the target rib for resection when the patient is obese or muscular. As a result, one or often many more intraoperative radiographs (usually cross-table lateral radiographs) with a radio opaque marker imbedded through the wound in the rib are necessary to verify the correct numbered rib to biopsy. Since it is not possible to obtain an intraoperative frozen section examination of the excised rib to verify that an abnormal area was excised, the final pathologic diagnosis awaits 7-10 days of decalcification and processing. And if the final diagnosis is "normal" bone, then there is more uncertainty whether the correct site was biopsied since some abnormality should be present to cause the localized uptake of radioisotope.

METHYLENE BLUE MARKING OF THE TARGET RIB

In view of the technical difficulty and uncertainty of conventional rib biopsy techniques using only the bone scan image as a guide, Little and associates in 1993 published their results of a technique they developed to mark the target rib preoperatively in the nuclear medicine department.[11] In this technique, the [99m]technetium diphosphonate radioisotope is injected 6-12 hours earlier and the patient is brought to the nuclear medicine department. An external technetium point source is positioned over the abnormal area imaged by the gamma camera. Once the point source is properly positioned and corresponds in several views to the abnormal area on the bone scan, the skin is marked with a pen. The area is anesthetized with lidocaine and a needle is inserted to the underlying rib where a small amount of radioisotope is injected locally to mark the rib. Further nuclear imaging is performed to verify that the area marked with the radioisotope corresponds to the hot spot on the bone scan. Several injections may be necessary to verify positioning. This is followed by an injection of methylene blue into the same needle left in place to stain the periosteum of the rib and the local soft tissue up to the skin. At this point, the patient is promptly taken to the operating room for an open biopsy of the blue-stained rib before the stain diffuses further into the surrounding tissues.

From this study, Little and associates[11] biopsied bones in 15 cancer patients (13 ribs, 1 skull and 1 scapula), and had a positive histologic result in all biopsies. However, only 8 of 15 (53%) had metastatic cancer and the other were benign conditions thereby yielding a false positive bone scan rate of 47%. Moores and collegues[12] in 1990 published a follow-up series of 33 bone biopsies using the same localization technique and a positive histologic abnormality was seen in 97% of biopsies. But once again only 52% of the entire series had a malignant focus found in the bone, so that the false positive rate was 48%. However in the 17 patients in this series with an abnormal scan but normal plain radiographs, the false positive rate was even higher at 71%.[12]

Although this methylene blue bone staining technique appears to be effective in experienced hands, it is somewhat time-consuming and cumbersome. The nuclear medicine department must carefully coordinate their time with an available operating room so that the patient may undergo immediate surgery after injection of the methylene blue to avoid spread of the dye to other ribs. In view of the considerable experience and coordination needed in the nuclear medicine department to allow smooth performance of this technique, the surgeon performing only an occasional bone biopsy using this method of localization might find it difficult to duplicate the excellent results of Little[11] and Moores.[12]

INTRAOPERATIVE GAMMA PROBE LOCALIZATION

The development of hand-held, high resolution gamma probes that provide real-time gamma counting intraoperatively has enabled the evolution of a much simpler, more direct technique to localize bone lesions for biopsy. These gamma radiation detection devices have been used quite successfully for intraoperative lymphatic mapping to locate and biopsy sentinel lymph nodes in melanoma[13] and breast cancer[14] patients. This technique has subsequently been adapted for use at our cancer center for the intraoperative localization of areas of increased uptake of radioisotope in ribs and sternum to guide the surgeon in open bone biopsy in patients suspected of having bone metastases.[3]

Technique

1. Three to twelve hours prior to surgery (preferably 4-6 hours for the strongest counts), each patient receives an intravenous injection of 28 mCi technetium 99m oxidronate, the standard dose for a radioisotopic bone scan.

2. General anesthesia is induced and the patient is positioned, prepped, and draped. The hand-held gamma probe in a sterile plastic sleeve is then used to locate the area of greatest tracer activity (measured in real time in counts per second) beginning in the anatomic area suspected from the bone scan. There is a moderate amount of background activity from the tracer in all of the nearby bones. The target "hot spot" on the bone scan when encountered by the probe will have a noticeably increased amount of tracer activity.

3. A 3-4 cm incision is made over the point of maximum activity and the bone is exposed down to the periosteum. Now the probe in the sterile sleeve is used to locate precisely the point of greatest activity in the surgical wound with comparison to the background counts on nearby ribs and elsewhere on the same rib. The point of greatest activity is then marked directly on the surface of the rib with the electrocautery just prior to removal of that section. Though this is a small wound, the tissues can easily be moved around enough to perform counting on the adjacent ribs immediately above and below the target ribs as well as further away from the hot spot on the same rib. The hot spot with its increased counts is usually very localized. In our series of biopsies of 18 ribs, the mean ratio of hot spot counts on the target rib to the background counts on adjacent ribs was 1.68. That is, the area of greatest tracer activity was an average 68% "hotter" by counts than the counts on surrounding bones, a readily discernible difference.

4. If desired, an intraoperative cross-table lateral chest radiograph may then be taken with a radio opaque marker on the rib (a spinal needle imbedded in the periosteum is my preference) to verify that the target rib is the correct numbered rib based on the bone scan. Usually, several radiographs are necessary to get the correct view so that the 1st or 12th rib is included in the image to enable counting of the ribs from above or below. After experience with only a few biopsies and after gaining

confidence in this radioguided technique, we eliminated this time-consuming and costly step.

5. A 3 cm portion of rib is then removed subperiosteally or the outer table of sternum or other bone is removed. The remaining ribs are then checked again in the wound with the gamma probe to verify that the hot spot target was removed. Of note, in the setting of a hot bone scan and normal rib detail plain radiographs, the resected piece of rib usually is grossly normal. The resected rib is placed in a solution to decalcify it and the specimen may be isolated for three to four days to allow the radioactive tracer to decay, depending upon the radiation safety requirements of your institution. Then, the rib is studied histologically with particular attention directed to the area of rib scored on the surface with the electrocautery.

6. The small wound is filled with saline to check to see if the pleural cavity was inadvertently entered, which occurred about 15% of the time in our experience. If the pleura was violated, a small 24 Fr chest tube is inserted through a separate skin incision and it is removed in the recovery room after the postoperative chest radiograph is shown to be normal (without a pneumothorax or pleural effusion). The wound is closed in layers with absorbable sutures, with a final subcuticular skin closure. Almost all patients are then discharged the same day from the recovery room, with only quite debilitated patients requiring observation overnight. A postoperative chest radiograph documents which numbered rib was biopsied.

Results

In our initial series[3] which is now updated and enlarged, 15 patients (9 men and 6 women) with a variety of known or suspected underlying primary cancers had positive radioisotopic bone scans but with normal plain radiographs, who were either asymptomatic or had minimal, nonlocalizing symptoms. A total of 19 bones (18 ribs and 1 sternum) were biopsied using this radioguided technique. All (100%) bones biopsied contained a pathologic process that accounted for the positive bone scan. However, only 4/19 (21%) of the bones demonstrated metastatic cancer (1 squamous cell carcinoma of the lung, 2 lymphoma, 1 carcinoma of the prostate). The other 15 bones (79% false positive rate for diagnosing metastatic cancer in the bone) showed a variety of benign pathologic processes: hypercellular marrow (4 ribs), fibrous dysplasia (1 rib), Paget's disease of bone (2 ribs), localized fibrosis with granulation tissue (1 rib), enchondroma (3 ribs), and chondroma (3 ribs and 1 sternum).

With the gamma probe directed biopsies, detection of the bone scan hot spot was easily accomplished. Intraoperatively, the bone for resection always appeared grossly normal to the surgeon. However, the mean ratio of measured hot spot activity on the target rib compared to adjacent ribs or the same rib away from the hot spot was 1.68 ± 0.32 (median 1.61, range 1.31-2.69), and the difference was easily discernible intraoperatively.

The most common benign abnormality in this series was a chondroma or enchondroma, accounting for 7/19 (37%) of biopsy results. This benign cartilaginous tumor is relatively common, representing 13.4% of all benign bone tumors although the actual incidence is unknown since they occur sporadically and are asymptomatic.[15] They tend to occur most commonly in the small bones of the hands and feet, but also occur in long thin bones such as the ribs. Unless they become very large, they remain asymptomatic and are not generally visible on plain bone radiographs. They are usually found incidentally as a hot spot on a bone scan performed during the evaluation of a patient with a malignancy. A subsequent biopsy usually follows and the diagnosis of a benign chondroma is made.

There was no morbidity or mortality specifically associated with the intraoperative gamma probe technique or the actual rib biopsies. With experience, the total operative time for these cases decreased to only 20-40 minutes. Intraoperative radiographs were eliminated because the counting technique is precise, rapid, and reproducible, making the radiograph unnecessary and/or redundant for localization of the abnormal section of bone.

CONCLUSIONS

The finding of an asymptomatic area of increased uptake on a radioisotopic bone scan with normal plain radiographs in a patient with a known or suspected malignancy does not necessarily indicate that bone metastases are present. Rather, there is a high false positive rate and as many as 75% or more will be benign lesions on biopsy. Histologic confirmation is mandatory in this setting.

Until recently, the surgeon faced with the prospect of accurately locating a bone lesion with only the bone scan image as a guide faced a real challenge. The technique of methylene blue "tattooing" of the target bone[11,12] appears to be effective but it is cumbersome and time consuming. Radioguided surgical techniques[3] offer real advantages in terms of decreased operative time, less interdepartmental coordination and 100% sensitivity. With the advent of multiple applications and widespread use of the gamma probe in melanoma and breast cancer surgery, this instrument is becoming a common fixture in most active operating rooms and it 14 will be available to guide surgeons in performing bone biopsies.

The gamma probe was used in our study[3] to guide the biopsy of asymptomatic rib and sternal lesions. However, this same technique could no doubt be easily adapted by orthopedic surgeons to biopsy subtle lesions in other bones in the appendicular skeleton. Likewise, the gamma probe could be used to aid in the biopsy of symptomatic and/or plain radiographically visible bone lesions when body habitus or bone lesion location might make precise conventional localization difficult intraoperatively. This technique has no side effects, it is easy to learn and apply, and should be strongly considered in the future for use by the surgeon in guiding the open biopsy of suspected asymptomatic bone metastases.

REFERENCES

1. Beahrs OH, Henson DE, Hutter RVP, Kennedy BJ. Handbook for staging of cancer. 1st ed. Philadelphia: JB Lippincott Co 1993:3-14.
2. Corcoran RJ, Thrall, JH, Kyle RW, Kaminski RJ, Johnson MC. Solitary abnormalities in bone scans of patients with extraosseous malignancies. Radiology 1976; 121:663-7.
3. Robinson LA, Preksto D, Muro-Cacho C, Hubbell DS. Intraoperative gamma probe-directed biopsy of asymptomatic suspected bone metastases. Ann Thorac Surg 1998; 65:1426-32.
4. Rogers LF. Secondary malignancies of bone. In: Juhl JH, Crummy AB, eds. Paul and Juhl's essentials of radiological imaging. 6th ed, Philadelphia: J.B. Lippincott Company 1993:164-5.
5. Clain A. Secondary malignant disease of bone. Brit J Cancer 1965; 19:15-29.
6. Henry JB, ed. Clinical diagnosis and management by laboratory methods. 19th ed. Philadelphia:WB Saunders Co 1996:277.
7. Michel F, Solèr M, Imhof E, Perruchoud AP. Initial staging of nonsmall cell lung cancer: Value of routine radioisotope bone scanning. Thorax 1991; 46:469-73.
8. Ichinose Y, Hara N, Ohta M et al. Preoperative examination to detect distant metastasis is not advocated for asymptomatic patients with Stages 1 and 2 nonsmall cell lung cancer. Chest 1989; 96:1104-9.
9. Wahner HW, Brown ML. Role of bone scanning. In: Sim FH, ed. Diagnosis and management of metastatic bone disease. 1st ed. New York: Raven Press 1988:51-67.
10. Ruchdeschel JC. Rapid cost-effective diagnosis of spinal cord compression due to cancer. Cancer Control 1995; 2:320-23.
11. Little AG, DeMeester TR, Kirchner PT et al. Guided biopsies of abnormalities on nuclear bone scans. J Thorac Cardiovasc Surg 1983; 85:396-403.
12. Moores DWO, Line B, Dziuban SW, Jr, McKneally MF. Nuclear scan-guided rib biopsy. J Thorac Cardiovasc Surg 1990; 90:620-1.
13. Albertini JJ, Cruse CW, Rapaport D et al. Intraoperative radiolympho-scintigraphy improves sentinel lymph node identification for patients with melanoma. Ann Surg 1996; 223:217-4.
14. Albertini JJ, Lyman GH, Cox C et al. Lymphatic mapping and sentinel node biopsy in the patient with breast cancer. J Amer Med Assoc 1996; 276:1818-22.
15. Unni KK. Chondroma. In: Unni KK. Dahlin's Bone tumors. 5th ed, Philadelphia: Lippincott-Raven 1996:25-45.

Fig. 1.3. (From page 13) Example of an intraoperative gamma probe, specifically designed and optimally engineered for radioguided surgical procedures, the Navigator gamma guidance system (US Surgical Corporation, Norwalk, CT).

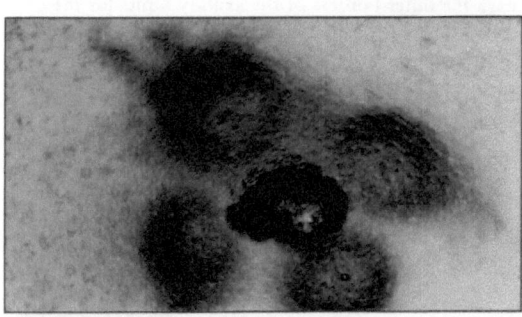

Fig. 5.3. (From page 67) Colored radiocolloid injection into the dermis of the skin in 4 quadrants around a nodular melanoma. Within minutes of the injection, tracer is seen being taken up by the cutaneous lymphatics. It is rare to have the primary lesion intact at the time of the mapping, and most of the data generated for success rates for melanoma mapping is after an excisional biopsy.

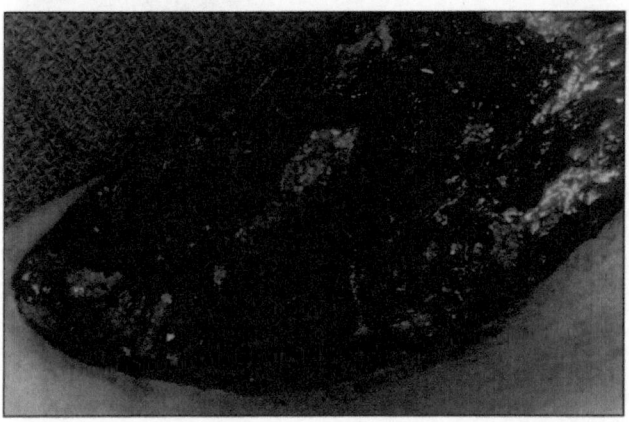

Fig. 5.4. (From page 67) A blue-stained afferent lymphatic is shown entering a blue-stained node. This node will be hot with the gamma probe, will be the SLN and will be the first site of metastatic disease.

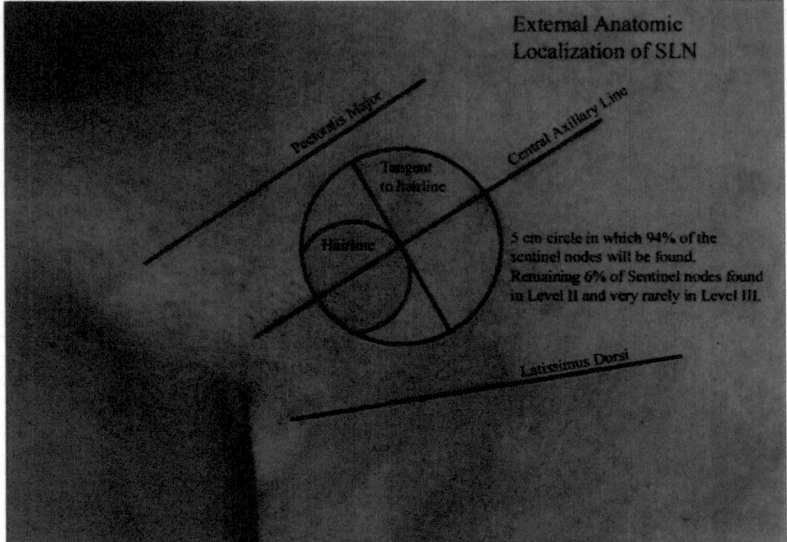

Fig. 6.1. (From page 78) Mapping the axillary SLN. If one were to draw a line along the lateral border of the pectoralis major muscle and lateral border of the latissimus dorsi muscle in the axilla, these would mark the outer borders of the axillary limits for the dissection. One should place a tangential line at the axillary hairline in a perpendicular fashion anterior to posterior. A line is then drawn through the axis of the axilla, through the center point of the hairline. Those intersecting lines mark the center of a 5 cm circle, which can be drawn on the axilla. Within this 5 cm circle are 94% of the SLN.

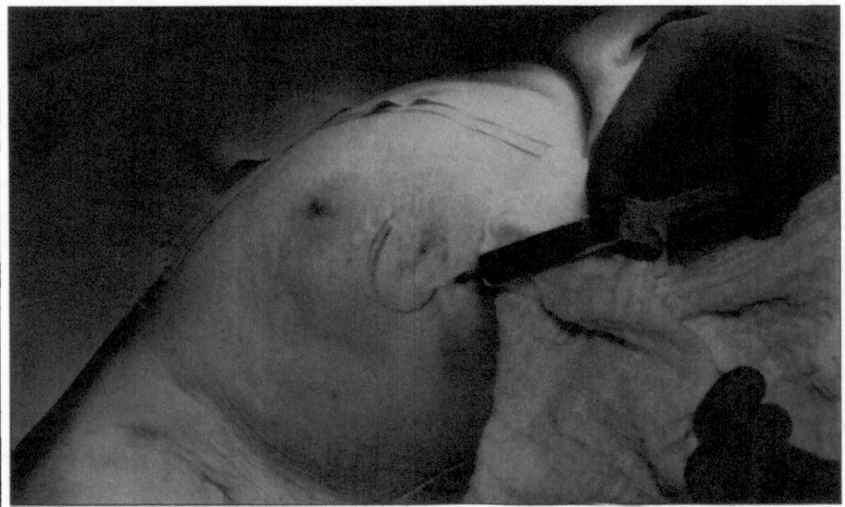

Fig. 7.1. (From page 84) Injection of isosulfan blue in the wall of the excisional biopsy cavity.

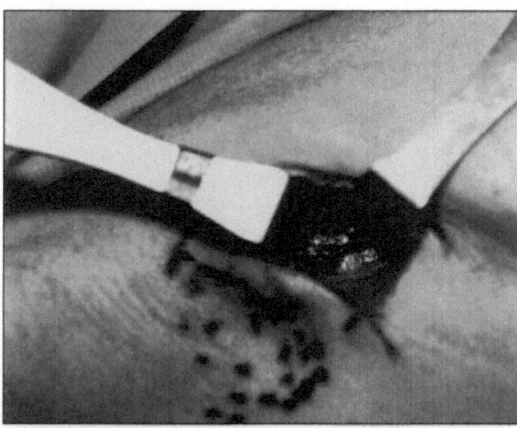

Fig. 7.2. (From page 85) Axillary incision about 1 cm below hair bearing area (indicated with purple markings).

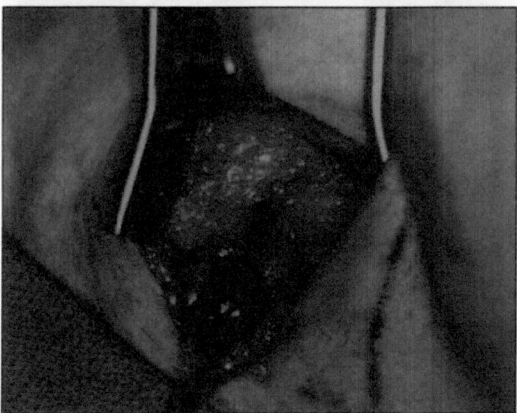

Fig. 7.3a. (From page 86) Blue lymphatic entering into blue-stained sentinel lymph node.

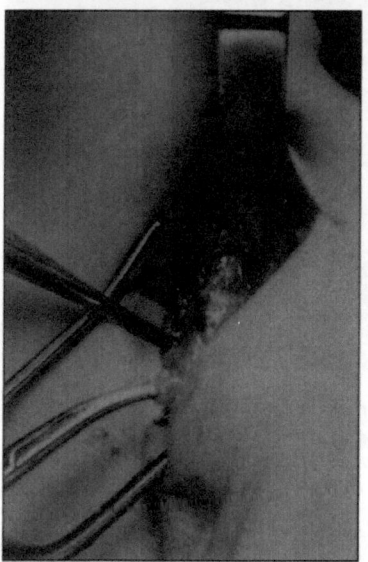

Fig. 7.3b. (From page 86) Another example of a blue sentinel lymph node in vivo.

Fig. 7.4. (From page 87) Excised blue
sentinel lymph node.

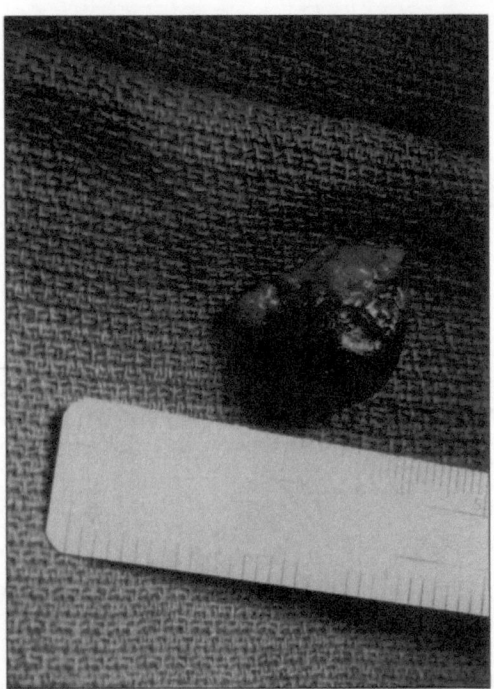

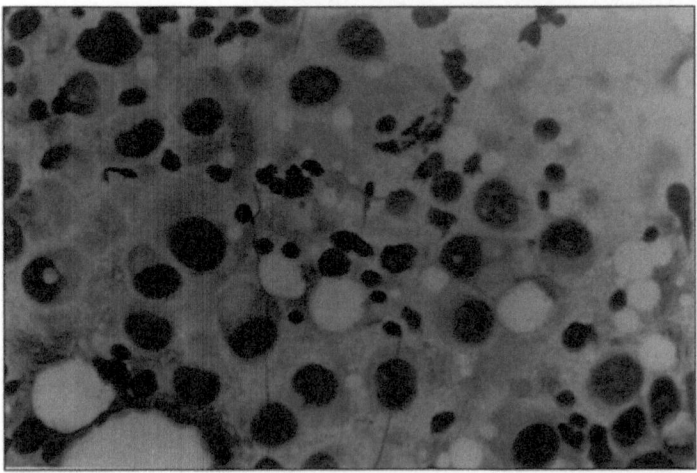

Fig. 9.1. (From page 109) Intraoperative imprint cytology of sentinel lymph
node with metastatic ductal Carcinoma as loose clusters and single cells, Diff-
Quik stain, X400

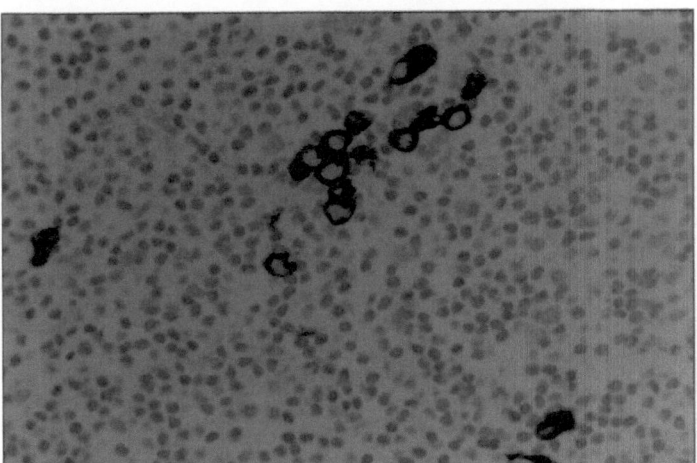

Fig. 9.2. (From page 109) Small microscopic tumor cells in single cell pattern distribution demonstrated only by cytokeratin immunostaining (Hematoxylin & Eosin stain negative), X400

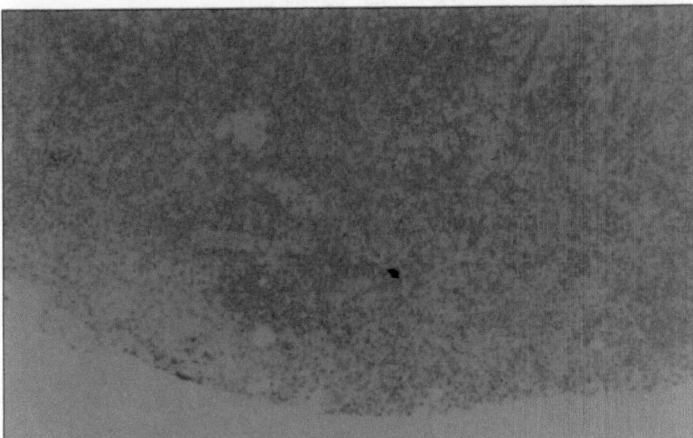

Fig. 9.3. (From page 110) Incomplete sectioning of the lymph nodal capsule missing the subcapsular micrometastases, hematoxylin & eosin stain, X100

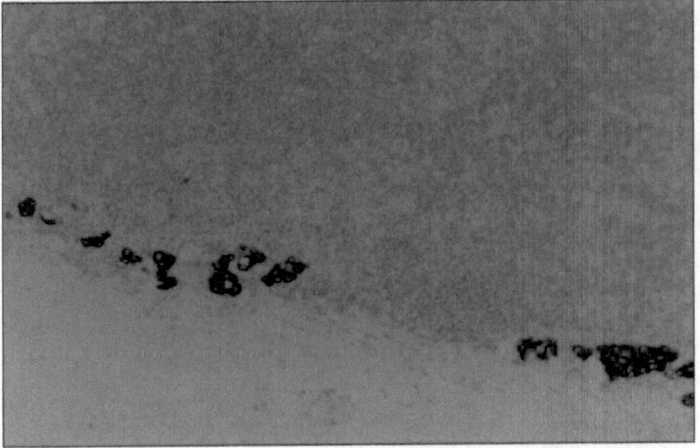

Fig. 9.4. (From page 110) Subcapsular micrometastases highlighted only cytokeratin immunostaining, X100 (deeper sections of Fig. 9.3)

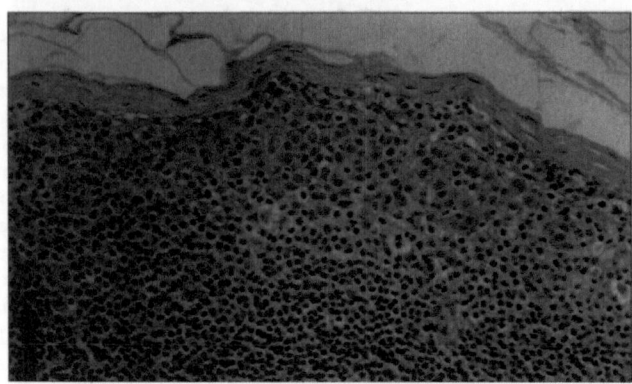

Fig. 10.1. (From page 118) Subcapsular aggregate of metastatic malignant melanoma.

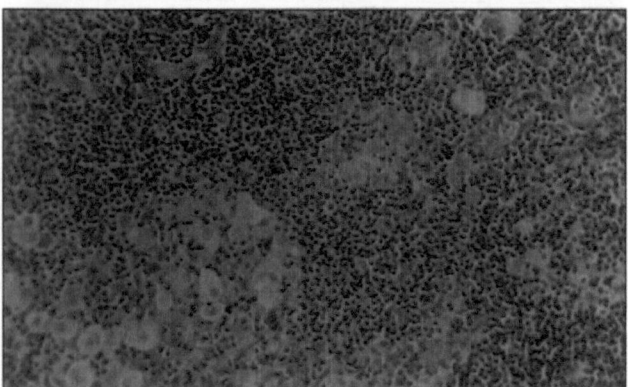

Fig. 10.2. (From page 118) Diffuse distribution of metastatic malignant melanoma.

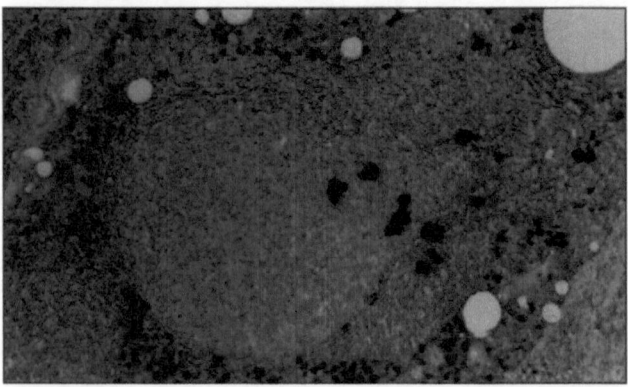

Fig. 10.3. (From page 119) Rare clumps of strongly S-100 positive cells of metastatic malignant melanoma, scattered weakly positive single dendritic cells.

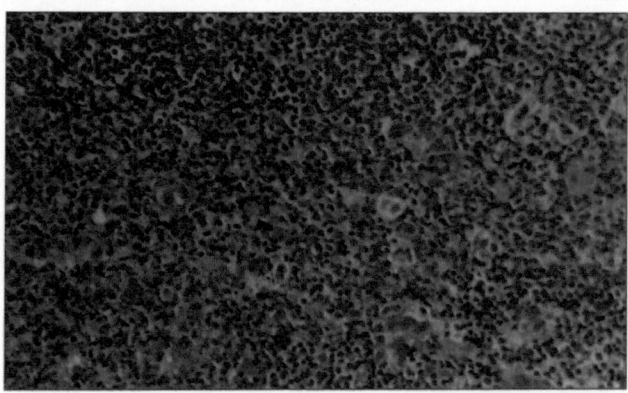

Fig. 10.4. (From page 119) Histologic section corresponding to Fig. 3, showing rare aggregates of pigmented, cytologically malignant cells.

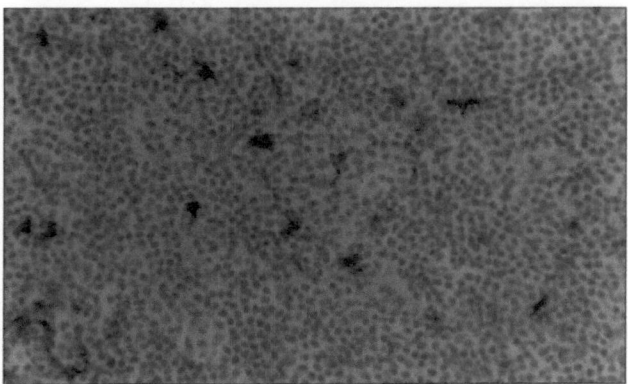

Fig. 10.5. (From page 121) Typical pattern of interdigitating reticulum cells.

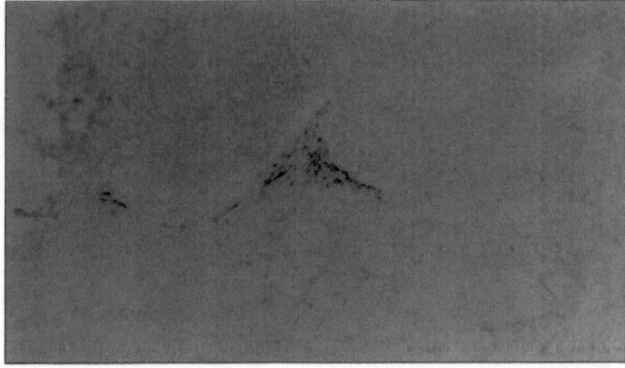

Fig. 10.6. (From page 122) Intracapsular aggregate of S-100 positive cells of a benign nevus cell aggregate.

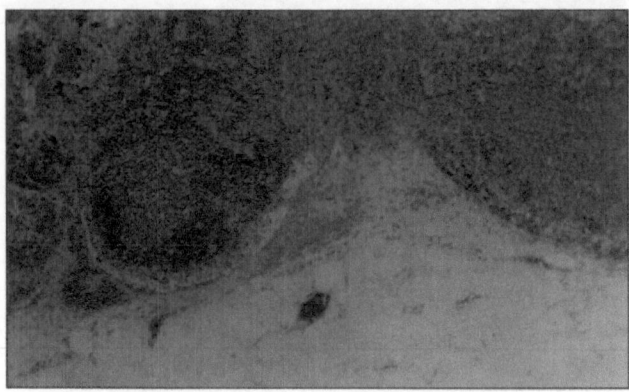

Fig. 10.7. (From page 122) Histologic section corresponding to Fig. 10.6, showing benign cytology of nevus cell aggregate.

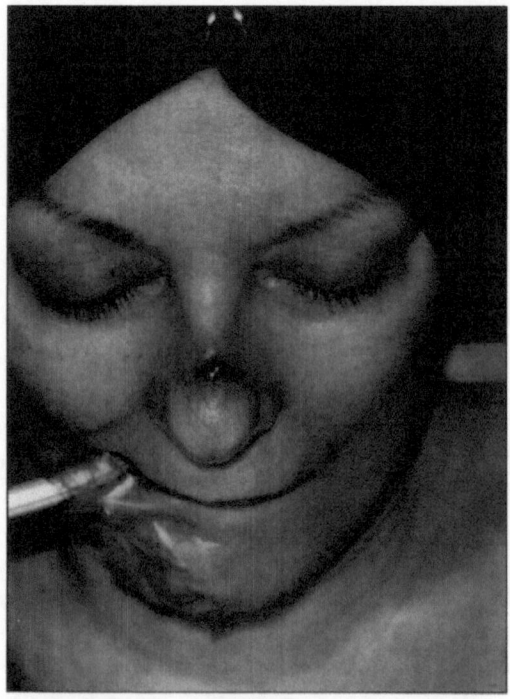

Fig. 12.3. (From page 138) 69 year old female with 4 x 5 cm MCC of nasal dorsum.